40-00

JAPAN'S MILITARY RENAISSANCE?

Japan's Military Renaissance?

Edited by

Ron Matthews
Senior Lecturer, School of Defence Management
Cranfield Institute of Technology

and

Keisuke Matsuyama
Associate Professor, University of Electro-
Communications, Tokyo

St. Martin's Press

Editorial matter and selection © Ron Matthews and Keisuke Matsuyama 1993
Text © The Macmillan Press Ltd 1993

First published in Great Britain 1993 by
THE MACMILLAN PRESS LTD
Houndmills, Basingstoke, Hampshire RG21 2XS
and London
Companies and representatives
throughout the world

A catalogue record for this book is available
from the British Library.

ISBN 0–333–57637–3

Printed in Great Britain by
Ipswich Book Company Ltd
Ipswich, Suffolk

First published in the United States of America 1993 by
Scholarly and Reference Division,
ST. MARTIN'S PRESS, INC.,
175 Fifth Avenue,
New York, N.Y. 10010

ISBN 0–312–09150–8

Library of Congress Cataloging-in-Publication Data
Japan's military renaissance? / Ron Matthews and Keisuke Matsuyama.
p.   cm.
Includes index.
ISBN 0–312–09150–8
1. Japan—Armed Forces.   I. Matthews, Ron.   II. Matsuyama,
Keisuke.
UA845.J39   1993
355'.00952—dc20
92–37814
CIP

'Japan is a great people. Her masons play with stone, her carpenters with wood, her smiths with iron and her artists with life, death and all the eye can take in. Mercifully, she has been denied the last touch of firmness in her character which would enable her to play with the whole round world.'

Rudyard Kipling, *From Sea to Sea*, 1889

Contents

List of Figures

List of Tables

x

Notes on the Contributors

Michael W. Chinworth is a Senior Analyst (Asia Technology) at The Analytical Sciences Corporation, Arlington, VA (USA). He has degrees from the University of Notre Dame, Indiana and the Johns Hopkins School of Advanced International Studies, Washington, DC. He has also studied at Seinan Gakuin University and Sophia University, Japan. Mr Chinworth was previously a Research Director of the Massachusetts Institute of Technology Japan Programme. He has written widely on Japanese and US international and security issues, and participated in the preparation of two influential Office of Technology Assessment Reports, *Arming Our Allies: Cooperation and Collaboration in Defense Technology* (1990) and *Global Arms Trade: Commerce in Advanced Military Technology and Weapons* (1991).

Alistair D. Edgar is a Graduate Fellow at the Centre for International Relations, Queen's University, Canada, and specialises in issues of defence procurement, industrial base policy and defence trade. His previous publications include works on West European collaboration in military aerospace projects, and on Canadian defence industries. He is currently preparing a study with David G. Haglund on the prospects for Canada's defence industrial base in the post-Cold War international defence market.

John E. Endicott is Professor of International Affairs and Director of The Centre for International Strategy, Technology and Policy at Georgia Institute of Technology (USA). He received his doctorate in international affairs from the Fletcher School of Law and Diplomacy, Tufts University. Prior to taking up his present appointment, Professor Endicott had a highly distinguished career in the United States Air Force, during which time he was Associate Dean of the National War College and Acting Director of the Institute for National Strategic Studies, National Defence University, Washington D.C. The Professor's research interests span Japanese security studies and American defence policy. He has published widely on these topics, including the following books, *Japan's Nuclear Option, The Politics of East Asia* and *American Defense Policy*.

Yutaka Fukuda holds dual appointments as an Associate Professor at the University of Electro-Communications and the University of Tokyo. His research interests are in Socio-economics and Socio-informatics.

Ian Gow is Deputy Principal at the University of Stirling (UK) and NatWest Professor of Contemporary Japanese Studies in the Department of Business and Management. He holds a diploma in Japanese from Osaka University and has conducted research and taught at Doshisha University and the elite Tokyo University. Professor Gow has also taught Japanese language and Japanese Business and Politics at the Universities of Sheffield and Aston before establishing the Japanese Business Policy Unit at Warwick University. He has written extensively on defence and technology policy matters, including *Japan's Quest for Comprehensive Security*, which has been translated into Japanese. He has given evidence to House of Lords' Committees on Japan–EC relations and technology matters, and is currently working on a major ESRC research project on Japanese Government–Industry relations.

David G. Haglund is the Director of the Centre for International Relations and Professor of Political Studies, Queen's University, Ontario, Canada. He received his Ph.D. from the Johns Hopkins School of Advanced International Studies, Washington D.C., and taught at the University of British Columbia prior to taking up his present appointment. Professor Haglund has been a NATO Fellow, and has been affiliated with the Institut des Hautes Etudes Europeennes, Strasbourg, France. He is the author of several books on defence-industrial issues, and is currently involved in research on the future of foreign troops in the new Germany, leading to a book, entitled *Homeward Bound? Allied Forces in the New Germany* (co-edited with Olaf Mager).

Keith Hartley is Professor of Economics and Director of IRISS and of the Centre for Defence Economics at the University of York (UK). He is joint editor of the journal *Defence Economics*, and Secretary-General of the International Defence Economics Association. He has written widely on defence economics and is the author of *The Economics of Defence, Disarmament and Peace* (1990).

Mitsuhiro Kojina is a Professor of Management Information Systems at Teikyo University. He has research interests in the relationship between informatics and economic development, and also the quantitative analysis of social systems. Professor Kojina has published numerous articles in this field.

Stephen Martin is a Research Fellow at the Centre for Defence Economics, University of York (UK). He holds a D.Phil. from the University of York and has published several articles on the economics of international collaboration. Dr Martin is currently engaged in a major ESRC-funded research project examining the implications of defence countertrade.

S. Javed Maswood (PhD Carleton University, Ottawa, Canada) is a Senior Lecturer in the Faculty of Asian and International Studies, Griffith University, Brisbane (Australia), where he lectures in international politics and international political economy. He has written numerous articles on Japanese political and security issues, and is the author of two recent books: *Japan and Protection* (1989) and *Japanese Defence* (1990).

Keisuke Matsuyama is an Associate Professor at the University of Electro-Communications, Tokyo. He holds several degrees in economics from Hitosubashi University. Professor Matsuyama's current teaching and research interests centre on econometrics with numerous articles on computational and economic cybernetics. He was recently a Visiting Research Fellow at the School of Defence Management, Cranfield Institute of Technology (UK).

Ron Matthews is a Senior Lecturer at the School of Defence Management, Cranfield Institute of Technology (UK). He has research interests in Third World industrial development and defence industrialisation. He has held several NATO fellowships, the prestigious Robert S. McNamara World Bank Fellowship, and has been a Visiting Fellow of: the Hoover Institute (US); the National University of Singapore; the Centre for Strategic and International Studies (Indonesia); the Institute of Strategic Studies (Pakistan); the Institute of Defence Studies (India); and the Institute of Development Studies (Kenya). He has published numerous articles on defence studies, and is the author of a recent book on *European Armaments Collaboration*, (1992).

Tomohisa Sakanaka is Professor of International Relations at Aoyama Gakuin University, Tokyo. Prior to beginning his academic career he worked for *Asahi Shimbun* (a Japanese national newspaper) as a Senior Staff Writer of Security Affairs with postings as Bureau Chief to Naha, Okinawa (when it was under the US Administration, 1962–4) and Saigon, South Vietnam (1975–85). Professor Sakanaka has been a Visiting Research Associate at the International Institute for Strategic Studies, London (UK). He has written many books and articles on Japanese and international security affairs.

Gerald Segal is a Senior Fellow (Asian Security) at the International Institute for Strategic Studies, London (UK). He was previously a Reader in international affairs at Bristol University, and prior to that taught at several other British Universities. Dr Segal has written widely on Asian political and security issues. His most recent publications include *Rethinking the Pacific* (1990) and *The Companion to World Affairs* (1991). Dr Segal is also the editor of the *Pacific Review*.

Trevor Taylor is Professor of International Relations and Head of the International Security Programme, Chatham House (UK). He has served on the Research Grants Board of the ESRC and is currently Vice-Chairman of the British International Studies Association. Professor Taylor has been awarded NATO Fellowships, and published numerous articles and books on defence-related issues, including *Defence, Technology and International Integration* (1982), (with Keith Hayward) *The UK Defence Industrial Base* (1989), and 'European Defence Industries: An Overview' (Jane's NATO Handbook 1991–92).

Akio Watanabe is Professor of International Relations at the University of Tokyo; a post he has held since 1978. He is a member of the boards of the Japan Association of International Relations, the *Japan Review*, and the Foundation for Advanced Information and Research. Professor Watanabe has been a Visiting Fellow at the Royal Institute of International Affairs, Chatham House (UK) and the Woodrow Wilson International

Centre for Scholars. He is also the author of innumerable articles and several major books, including: *The Okinawa Problem: A Chapter in Japan–US Relations* (1970); *The Foreign Policy of Modern Japan* (1977); and *Government and Politics in Modern Japan* (1989).

Preface

In political and military terms 1992 was a watershed year for the Japanese. In the June of that year Japan's upper house of Parliament approved a bill allowing Japanese troops to serve abroad. In late 1992, a small group of 30 Japanese army engineers were despatched to Cambodia, making it the first time since late Second World War that Japan's military has served overseas. For Japan, with its famous Peace Constitution, this was indeed a significant step, and raises important questions for the international community. For instance, does this policy 'sea-change', which allows up to 2000 Japanese soldiers to participate in United Nations peacekeeping operations, represent the thin end of the wedge for Japanese remilitarisation? Furthermore, will the easing of the self-imposed constraints on Japan's military legitimise the country's expansion of defence production and military capability that has taken place unabated throughout the 1980s and early 1990s?

In the event, Japan's peacekeeping forces will be remarkably unmilitary: unarmed except for pistols for self-defence, they will be technically required to withdraw from the scene of combat if the UN forces to which they belong come under fire, and will not even be allowed to take part in mine-clearing operations. The breaking of this postwar taboo on Japanese overseas military operations has nevertheless aroused the latent suspicions of Japan's nervous Asian neighbours. Korea, Taiwan and China have all publically expressed concern over Japan's increased military profile. Indeed, it is an open secret that China, in particular, sees 'pacifist' Japan as a major threat. Fifty years has apparently done little to erase Asia's agony of Japanese imperialism.

In the short run Japan poses no military threat: its rise in defence expenditure is now waning; military capability is primarily defensive in character; and the authorities, conscious of international susceptibilities, are active in assisting multilateral initiatives to promote peace. No, the short run is not the problem, it is rather the unpredictability of the longer term which attracts concern.

This book has two principal aims. Firstly, it seeks to be

informative. It brings up to date a subject-area which the literature has largely confined to the history books of the Pacific war. Much of the information provided by this book's authors will be of interest and even surprise informed Japanologists, including academics, politicians and strategists. However, given the international clamour for all things Japanese, the book is also intended to appeal to the broad cross-section of general readers who have a keen interest in international affairs. The book's second aim is to offer, through a spectrum of international perspectives on related but different politico-economic and social topics, a critical evaluation of the issues surrounding Japanese militarisation, and to speculate on the future direction it will probably take. The intention is to open debate, rather than settle it, on a subject that will be of increasing interest and controversy throughout this decade and beyond into the second millenium.

This book would not have been possible without the cooperation and commitment of all its contributors. To coordinate and schedule this type of international project is no easy task. In the event, however, it was a smooth, uncomplicated operation due to the energy and enthusiasm of the authors. The seeds of the project were germinated in 1991 during lively lunchtime discussions between Visiting Japanese Professor Keisuke Matsuyama and myself at the School of Defence Management, Cranfield Institute of Technology, England. My co-editor handled the administration and coordination of the project from the Japanese end, while I was responsible for editing and coordinating western inputs. From both the editorial and literary side, the usual disclaimers apply.

We acknowledge the support of all those, known and unknown, who have assisted in the publication of this book, in particular, Belinda Holdsworth, our publisher, for her patience, Beryl Bell and my wife, Jenny, for their typing of manuscripts, Lt-Col. Jeremy Bethell for his accomplished editorial and computer skills, and Air Commodore Peter Markey OBE and Ann Vu for specialist advice. The final acknowledgement goes to my sons, Tommy and Toby, who bore the absence of their father over weekends and evenings with remarkably good humour.

RON MATTHEWS

1 Japan's Security into the 1990s
Ron Matthews[1]

I INTRODUCTION

Power is a multifaceted concept. There are, for example, economic, political and military dimensions of power. Although each of these power attributes can exist in isolation, the reality is that they feed off each other. Historically there is correspondence between politico-economic power and military strength. Modern Japan, with its well-known 'peace' Constitution, appears the exception. It is the world's second most powerful nation. Yet it forswears the possession of 'offensive' military capability, constrains itself to defence expenditure at around 1 per cent of national income, and effectively bans arms exports. For a powerful nation this is indeed unique.

The purpose of this book is to explore Japan's military status. Japan's military is already among the most powerful in the Asia-Pacific region. As the next century beckons, and Japan moves inexorably towards becoming the world's most powerful economy, will resulting domestic and international pressures elevate Japan's military profile? Is it inevitable? Is it something to be feared? These are important issues. They need to be raised and indeed addressed, as with the passage of time they become both more urgent and more controversial.

As a departure point, this opening chapter provides the backdrop for exploring the implications of Japan's postwar defence allergy in the face of increasing pressures to rearm.

Victor or Vanquished?

There are remarkable similarities in the international order between the 1930s and today. The same three economic power houses, the United States, Germany and Japan, dominate international trading and increasingly political systems. Due both to its economic and military muscle, the United States retains its

1

superpower mantle. In Europe a reunified Germany wields pervasive influence, even given the temporary 'blip' of economic disruption flowing from reunification. The disintegration of the Soviet Union has acted to reinforce Germany's relative power dominance. And in the Asia-Pacific region Japan has re-emerged as the centrifugal force radiating economic influence throughout the area. Asian development, for instance, has largely been leveraged by Japanese direct investment.

For many Western observers the postwar politico-economic rehabilitation of Germany is acceptable; indeed welcomed, but Japan's transformation process has been less well received. The Japanese people's mesh of deference and superiority bemuses the occidental. Trading frictions in particular have long characterised Japan's relations with the West. Unlike Germany, Japan's success has never easily been accommodated by the Western nations. 'Power' appears to pose a problem. The reasons for this are diverse. They include: Japan's alleged militarist genealogy; its supposed subliminal pursuit of a 'divine mission' in Asia; and classical mercantilist leanings coupled with a growing economic assertiveness. The ultimate goal, it is commonly held, is world 'economic control'. Moreover, Western fears in this respect have been heightened in the 1990s by a growing recognition that technological and commercial capabilities more than military strength are the significant determinants of state power and influence. A January 1992 comment by Mr Sakurauchi, speaker of the Lower House of Japan's Diet, highlights the nationalistic overtones of the global 'economic race' when he described the US as Japan's subcontractor.[2]

Western sensitivities are particularly acute over the role of Japan's military. There has been a barrage of criticism directed towards Japan's economic success, suggesting that it has been achieved through 'free-riding' under the US nuclear umbrella. An additional related criticism has been Japan's avoidance of international responsibilities commensurate with its status as a world power. Yet the dilemma facing the Japanese is that they are damned if they do, and damned if they don't. The lack of a Japanese military contribution to the Iraq–Kuwait conflict was widely condemned at the time, especially given that Japan receives 60 per cent of its oil via the Gulf of Hormuz compared to only 7 per cent for the United States.[3] But Japanese troops in active combat would certainly have provoked greater controversy. Moreover, much of the angst undoubtedly would have been sown from those nations active in encouraging Japan's 'rearmament'. The fact is that Japanese militarism evokes a chilling stereotype. It has caused what has been termed a 'legitimacy deficit', where Japan's rightful claim to international leadership, based on economic strength is judged on historical rather than contemporary considerations.[4]

This is a rather strange state of affairs, however. After all, Germany has been accepted back into the international fold, and is an important member of NATO. Moreover, current Japanese militarisation (rearmament) does not necessarily connote militarism (malevolent intent supported by armed force). Japan's postwar democratic mechanisms and institutional checks assuage the resurgence of an omnipotent military establishment. There is thus little need to fear a causal relationship running from militarisation to militarism. Nevertheless, the visceral urge persists. One plausible explanation for the longevity of the 'yellow peril' syndrome may be because Japan's culture, religion and social fabric are different from other countries, including those Asiatic. Consequently, it is difficult to feel secure about Japanese thinking and intentions. The ambiguities and contradictions between Japan's constitutional position on defence and what happens in practice encourages misinterpretation. The cause is the unique conundrum Japanese policy-makers face: how at US behest to sponsor an increased military profile while at the same time maintaining the credibility of a benign foreign policy image.

A useful framework for examining Japan's response to this dilemma is via the notion of 'comprehensive security'. Introduced in the 1980 Report of the Comprehensive National Security Study Group, the logic of the approach rests on the understanding that security means more than military might. Due to Japan's dependence on imported food, raw materials and energy, security necessarily encompasses economic as well as military elements, each of which complements and reinforces the contribution of the other. If Japan in the 1990s is to adopt a consummate approach to comprehensive security then the political role of international diplomacy must also be added. The interfaces between these various forms of security are as blurred as the concept of 'comprehensive security' is itself indisputably vague. Yet while this approach is not formally enshrined in

Japanese policy, it does capture the broader sense of the challenges facing Japan. Firstly, regarding the fluidity of the geo-strategic environment, there are the political uncertainties following the break-up of the erstwhile Soviet Union, undermining the stability and predictability of strategic relationships and planning in Northeast Asia. But insecurities extend further than this. Japan, more than most Asiatic nations, suffers economic vulnerabilities which are brought to the fore during times of uncertainty. Note here, for instance, the uncontained and recurrent military threats from the Middle East, Indochina, and Southeast Asia, threatening secure and continuous passage of vital oil, food and material supplies to Japan. Secondly, there are the moral and strategic imperatives of humanitarian aid and development assistance to needy countries, cultivating binding long-term relationships with aid-recipient states. Finally, there is the need for Japan to exert influence in, both extant and prospective, international policy fora on matters of interest to Japan. Membership of the United Nations (UN) Security Council provides one relevant example here.

II EVOLVING MILITARY POSTURE

Japan has changed irrevocably since the Second World War. Economically, socially and even culturally the country has moved on. At first sight the role of Japan's military establishment has been left untouched by myriad developments affecting the country. This is both true and untrue. There is no doubting that the perceived unimportance of the military in Japan has remained a constant over the years. For while there have recently been a number of subtle changes in defence policy, the foundation of Japan's defence posture – the 1947 Constitution – remains unaltered.

Article 9 of General MacArthur's peace Constitution has to be the most widely quoted clause of any country's constitution. It comprises two main parts: the first pertaining to war renunciation and the second to the prohibition of 'offensive' war potential.

The Constitution has never been amended. Hence the existence of around 250 000 Japanese military personnel surely hints at something unconstitutional. In the Japanese manner, it is all a

matter of interpretation. In understanding this viewpoint there
is a need to explore the origins and controversies of the 'peace'
clause. Aimed at preventing Japan from ever again waging war,
Article 9 was modelled on the international Kellogg-Briand Pact
of 1928. The purpose of the Pact signed and ratified by the
United States, the former Soviet Union, the United Kingdom,
France, Germany, Japan and many other nations was to ban
war. An interesting feature about this Agreement apart from its
total ineffectiveness has regard to two conditions added by the
US Senate. These were (1) that any nation would be allowed to
defend itself by force if attacked and (2) that each nation would
decide for itself what constituted self-defence. Notwithstanding
the view that the second clause, in particular, apparently per-
mits all but self-declared wars of aggression, of which history
records few, if any,[5] Japan's Article 9 seems to mirror both the
principal intention and loose wording of the Kellogg-Briand
Pact. Japan was no longer to be able to wage war, although
'self-defence' was an inalienable right; the Japanese interpretation at least.

Japan's official interpretation of Article 9, which became *com-
munis opinio doctorum*, is that it retains a right of national self-
defence in international law but could neither wage war nor
maintain an armed force – seemingly, even for the purpose of
self-defence. For pragmatic purposes, attempts to overcome this
constitutional paradox have led Japanese government officials to
perform diplomatic gymnastics in the postwar period. Efforts to
justify the existence of Japan's Armed Forces have attempted to
deflect attention from the Constitution, focusing instead on Article 51 of the United Nations Charter, which states that self-
defence is a right of every signatory nation.

It was the outbreak of the Korean war in 1950 that eventually
led to the creation of Japan's postwar Armed Forces, or Self-
Defense Forces as they came to be called. This development
represented the beginning of what a recent Japanese prime
minister, Yashiro Nakasone, describes as a transition from a
'peace country' (*heiwa kokka*) to an 'ordinary country' (*zairaigata
kokka*).[6] Between 1945 and the outbreak of the Korean conflict
Japan possessed no armed forces. Thereafter, to replace Japan-
based US forces deployed in Korea the Americans encouraged
the establishment of a 75 000 Japanese Police Reserve Force.
Members of Japan's old imperial army joined this force, which to

some observers was a mere disguise for the organisation of a new Japanese army.[7] This Police Reserve, comprising ground and maritime forces, had evolved by April 1952 into something called the Japanese National Safety Forces (NSF). Prime Minister Yoshida who acted as the 'safety' Minister had to be intellectually nimble-footed in response to awkward questions about his portfolio. Against opposition party attack, Yoshida maintained that war potential forbidden by Article 9 could be differentiated from 'defence potential'; also that the NSF were not unconstitutional because they had no capability to wage modern warfare, and thus were not an offensive threat.[8] Yoshida later gave the example of a jet aeroplane, something which Japan did not possess at that time, as constituting war potential. This definitional dispute over defensive vs offensive military capability represented the opening shots of a debate that has continued to rage. In practical terms the controversy manifested itself in, for example, the removal of bombsights and inflight refuelling devices from Japan's F-4EJ Phantoms in the 1960s. These items were interpreted as offensive in character, although inexplicably not being viewed as such when F-15J fighters were introduced in the 1980s.

The limitation that a solely defensive capability imposes on Japanese security is compensated by the country's reliance on the United States–Japan Defence Treaty. The original Treaty, signed in 1951 (at the same time as the San Francisco Peace Treaty), was tantamount to the continued occupation of Japan by a foreign power. In 1960 a revised defence treaty came into force. In substance this provided a potential aggressor with the foreknowledge that either a nuclear assault or a large-scale conventional attack on Japan would risk direct confrontation with the United States. For the United States, the treaty's major benefit was Japan's formal integration into the strategic chain of countries blocking expansion of communism. To Japan, by contrast, the treaty meant that the authorities could concentrate attention and resources on economic rehabilitation unencumbered by military build-ups.

Defence was not ignored entirely, however. Over time there has been a gradual, more militaristic interpretation of the peace Constitution, which its ambiguity allows. In 1954 the Japanese Defense Agency (JDA), succeeding the National Safety Agency, was established to oversee the Ground, Air and Maritime Self

Defense Forces (SDF). In 1956 the National Defense Council was created with a mandate for formulating the size and composition of the SDF. A year later saw the adoption in Cabinet of the 'Basic Policies for National Defense', the fountain-head for all subsequent defence programmes. In 1957 Japan's first Five Year Defense Build-up Plan was endorsed providing for the expansion and modernisation of the country's SDF. Later in 1976 greater flexibility was introduced through the publication of a document entitled 'National Defense Program Outline'. Here, two important policy changes were signalled: firstly, the hitherto approach of planning to meet a specific threat was discarded in preference for a defence structure smoothly adapted to confront emergency situations. Secondly, the previous pattern of fixed build-up programmes spanning a given period of time was to be scrapped and replaced with a more flexible system where decisions would be taken each fiscal year: this annual approach was short-lived, however. The principle of strategic planning was reinstated in 1980 allowing defence planning to again adopt a longer-term perspective. This was due to the perceived danger that flexibility might encourage 'creeping' militarism. But although the switch away from fixed programmes proved temporary, it did have one effect: it forced Prime Minister Miki, under opposition concern, to limit defence spending to 1 per cent of GNP 'for the time being'.

In the event the life of the 1 per cent ceiling was short. In 1987 Prime Minister Nakasone while allaying fears over the reincarnation of Japanese militarism conceded that for the Five Year Defense Plan (1986–90) the 1 per cent threshold would be broken. Justifying this policy-break he argued that 'while [the plan's raised expenditure] comes to 1.02 per cent of GNP on a yearly basis, it is almost the same as 1 per cent'.[9] Notwithstanding the Prime Minister's terminological impreciseness the 1 per cent threshold had been broken. And of course if items excluded from Japan's defence budget are taken into consideration as they are in calculations by NATO countries then Japan would have consistently broken through the 1 per cent barrier. It should be further noted that Japan's defence expenditure to GNP had regularly been above 1 per cent over the period 1955–65. It was due to Japan's GNP growth outstripping the rise in defence expenditure over the 1970s and early 1980s that the ratio fell. This is not to say, however, that Japan's defence spending declined. In fact the opposite occurred: in absolute terms military expenditure has risen continuously since the inception of the JDA.

Export policy was similarly restrictive. Japan's policy stance here was established in 1969, and over time has effectively led to a ban on arms exports. It should be noted, though, that this ban is not strictly a ban: Japanese policy officially stating that exports of certain items on export control lists will be 'restrained'. In scope the government 'banned' exports of firearms, ammunition, military vehicles, ships and aircraft. The regulations were further strengthened in 1976 to include similar restraining controls on the export of weapons manufacturing facilities.[10]

During the last decade, political and economic pressures have led to the circumvention of these restrictions. To begin, Japan's restrictive policy on military exports was officially broken in 1983 by the provision of a unique exception, the United States. In 1986 Japan announced its approval of the first transfer of military technology relating to KEIKO-SAM missile information. This agreement was legal. Disconcertingly, however, reports of illegal weapons exports have surfaced from time to time: for example, sales of Japanese mines and bombs to India and Taiwan.[11] There were also disclosures in 1991 of exports of Sidewinder missile technology and navigation equipment for F4 Phantom Jets to Iran. US–Japan relations were strained over this affair, with the United States particularly worried about military technology leakage.

There is also the intractable issue of dual-use items (resources applicable to both civil and military activities, although not readily identifiable as war material) which have for many years been exported by Japan. Professor Taylor examines this issue in detail (see Chapter 11), but it is instructive at this point to gain a sense of the scale of such trade. Japanese machine tools, jeeps, helicopters, chemicals and patrol boats have all found their way to the Middle East, Africa and the Far East. Also for years the BK 117 multipurpose helicopter jointly produced by Kawasaki Heavy Industries and Messerschmitt-Bolkow-Blohm GmbH has been used by the German military. In addition, exports of military-related equipment to communist countries have breached COCOM export controls to communist countries on several occasions. The Toshiba Kongsberg affair involving the export of strategically sensitive military machinery to the Soviet

Union in 1976 is notable here. Furthermore, as *The Economist* observes, the export surge in China's arms business (ranked number five in the world, with US $926 million worth of military exports in 1990) owes as much to Japanese parts and components as it does to China's recent military cooperation with the US.[12] But the united States itself benefits from Japanese exports of dual-use technology, especially semiconductors. Here the degree of penetration into America's military market has recently been the subject of controversy. Japanese authors Shintaro Ishihara and Akio Morita (in a book entitled, *The Japan That Can Say 'No'*) created a furore in the United States when they argued that Japan held the balance of power because of the Pentagon's dependence on Japanese high technology imports. The debate was further fuelled by press comment during the Gulf War that US advanced war planes, cruise missiles and 'smart' weaponry were only operational because of Japanese software. Although the claims are probably exaggerated, they nevertheless raise two important issues. Firstly, when Toshiba or NEC sell, say, semiconductors to General Electric, these Japanese companies are unlikely to be aware of the ultimate product destination, whether it be the American conglomerate's consumer, professional or military products.[13] Secondly, although much of the critical chips supplied for use in sophisticated military applications are probably sourced from American *niche* suppliers, there is no doubting that the levels of Japanese dual-use technology sales in the US market is attracting increasing official concern; an issue we shall revisit in the next section.

Not only does Japan prohibit foreign sales of military products and process technologies but also the overseas deployment of military personnel. This statutory restriction dates back to the SDF establishment law in the 1950s. However, the political fallout from the Gulf crisis led to a questioning of the role of Japan's SDF. Recently Parliament has debated to overturn the 1954 resolution prohibiting the dispatch of Japanese troops abroad for any purpose. The $13 billion payout to the Alliance combatants in the Gulf War was never likely to completely placate concerned Western countries nor nationalists within Japan. But although the Liberal Democratic government's initiative to establish up to 2000 troops armed only for self-defence for UN peacekeeping operations (once ceasefires are in place) has faced difficulty in earning parliamentary approval, it clearly represents an important policy *démarche*. An equally significant development, however, was Japan's deployment to the Gulf, albeit after the war had ended, of four mine counter-measure vehicles plus two support ships. Interestingly this follows the precedent set during the Korean war when Japan's minesweeping fleet (the biggest in East Asia at that time) was instrumental in effecting American landings on the Korean Peninsula.

Japan's self-imposed military restrictions also extend to non-conventional defence. These derive from the government's 1967 enunciation of the three non-nuclear principles whereby Japan pledged not to possess, produce or allow entry of nuclear weapons into Japan. Japan has repeatedly reaffirmed adherence to these three principles. Testimony to this fact comes from Japan's participation in the nuclear Non-Proliferation Treaty, which it signed in 1976. In addition, a legal barrier to nuclear armament is embodied in Article 2 of Japan's Atomic Energy Act confining research, development work and application in the field of nuclear energy to peaceful uses of the atom. The pursuance of an extremely large civil nuclear programme (Japan has one of the world's most extensive civil nuclear energy networks comprising 39 nuclear power-plants in operation with a further 11 under construction and three more planned)[14] is not inconsistent with the disavowal of a military nuclear capability. However such a programme does create for Japan much of the infrastructure and resources necessary for exercising a nuclear option in the future should occasion warrant it. Indeed, fears of Japan acquiring nuclear weapons have occasionally surfaced, as evidenced for instance by the March 1992 leaked Pentagon Defense Planning Guidance document. Significant in this regard is the statement by Prime minister Kishi in March 1959: 'The Government intends to maintain no nuclear weapons, but speaking in terms of legal interpretation of the Constitution there is nothing to prevent the maintaining of the minimum amount of nuclear weapons for self-defense.'[15] In addition, the third non-nuclear principle – no entry of nuclear weapons into Japan – lacks credibility. To date, Japan has never raised questions concerning the US navy's 'neither confirm nor deny' policy on whether US warships which visit Japanese ports are armed with nuclear weapons.

The government has also indicated that space will be used only for peaceful purposes.[16] Here again, however, the impartial observer is left pondering Japan's interpretation of 'peaceful purposes'. The US Strategic Defense Initiative (SDI) is a military project. And since Japan's decision in September 1986 to join in the Ronald Reagan Administration's SDI some of the country's biggest defence-industrial contractors such as Mitsubishi, Kawasaki and Ishikawajima-Harima are participating in this military-driven space programme.

The nuances of Japan's constitutional and policy posture have afforded it an increasingly flexible interpretation of self-defence. Over the last decade the JDA has enjoyed one of the fastest defence expenditure growth rates in the world, averaging 6.5 per cent between 1980 and 1989. Defence expenditure for fiscal year 1992 approximates US $36 billion, rising by around 3 per cent over the 1991–5 Mid-Term Defense Build-Up Plan. Military spending at this rate will just keep below the 1 per cent ceiling. Although the growth of defence expenditure is under review the fact remains that Japan alone among the Group of Seven industrialised nations continues to increase military spending. The size of Japan's economy coupled with the fragmentation and economic demise of the former Soviet Union suggest that *ceteris paribus* Japan's defence expenditure could soon be the second highest in the world. This is certainly an issue of some significance. Japan's military dominance is particularly telling in the Asia-Pacific region. As Table 1.1 shows, Japan's relative defence expenditure is substantial: five times that of Australia, three times that of North and South Korea combined, and greater than the total military expenditure of all the countries listed. Although not shown, Japan's defence spending is also nearly five times that of the ASEAN countries combined and almost 20 per cent greater than China's.[17]

The high absolute value of Japanese defence expenditure is reflected in the country's build-up of the SDF's military capability. The Japanese Air Self Defense Force (JASDF) possesses 46 000 personnel. The 130 F-4EJ Interceptors are in the process of modernisation. These and the other fighters including the advanced F-15EJ Eagle total about 340 planes, which is similar to the number the US Air Force has defending the continental United States.[18] The JASDF also has at its disposal some of the West's most advanced air defence systems including the Patriot surface-to-air missile that proved so effective in the Gulf conflict. In addition there are eight Grumman E-2c Hawkeye early

Table 1.1   Asia-Pacific defence expenditure (1990)

Country	Defence expenditure ($mn)
Australia	5 951
Indonesia	1 700
Japan	30 483
Malaysia	1 884
N. Korea	2 003
S. Korea	1 827
Philippines	676
Singapore	1 433
Taiwan	6 562
Thailand	2 392

Note: Values quoted in US $mn, at 1988 prices and exchange rates.

Source: *SIPRI Yearbook 1991: World Armaments and Disarmament* (SIPRI, Stockholm, 1991) table 5A.2.

warning aircraft with a further five on order. Also, of course, on the horizon is Japan's sophisticated FSX fighter plane.

The strength of the Japanese Maritime Self Defense Force (JMSDF) is equally impressive. In size it is bigger than Britain's Royal Navy. The 44 000 naval personnel man a fleet of 16 submarines and over 60 surface ships including 42 destroyers and 16 frigates. Not only does this represent one of the most modern navies in the world in terms of hull life, but in numbers it is nearly three times the size of the US Seventh Fleet whose strategic responsibility covers the Western Pacific and Indian Ocean. The JMSDF's 100 P-3c antisubmarine warfare aircraft is four times that deployed by the US Seventh Fleet.[19] More significant is the construction by Ishikawajima Harima of a 5500 ton dock landing ship which has the potential for operating VSTOL aircraft. This could be viewed as an interim step towards the JMSDF acquiring an aircraft carrier (an issue prone to intermittent speculation). Although there is no inevitability to this development, events, as Gerald Segal rightly relates (Chapter 6), turn very much on the nature of the strategic environment facing Japan. But it is important to note that Japanese fighters are operating at their range limits since 'self defence' was reinterpreted in the mid-1980s to incorporate a maritime defence zone expanded from 200 to 1000 nautical miles southeast of Honshu.[20]

The final branch of the Armed Services, the Japanese Ground Self Defense Force (JGSDF), is 160 000 strong with 44 000 personnel in reserve. It has 1200 tanks, soon to include 200 of Japan's indigenously developed Type-90 model and 930 armoured vehicles. The JGSDF is also being armed with the locally developed SSM-1, a surface-to-surface missile designed to attack enemy shipping using the straits and approaches contiguous to Japan.

III   REGIONAL THREATS

Japan's growing military strength behoves the question, why? The simple answer is that the military threat in the Asia-Pacific region remains undiminished. In 1990 the region's defence expenditure topped $86 billion, the highest regional total outside the NATO area. Moreover, it continues to grow. Based on threat perception and counter-perception an arms race is under way. For this region, institutionalised arms agreements do not exist and the 'peace dividend' is still some way off. In consequence the region remains tense. Northeast Asia, in particular, is still caught between the Scylla of Cold War confrontation and the Charybdis of recent developments.[21] Although the Soviet Union has disintegrated, the military threat remains substantial. According to Japan's 1990 Defense White Paper, Soviet Far East reinforcement and service upgrading includes deployment of 'Oscar II' nuclear-powered cruise missile submarines, *Slava* class cruisers, T-80 Main Battle Tanks and TU-26 'Backfire' bombers. Thus even though the size of the erstwhile Soviet Union's eastern military capability is shrinking, its quality is rising. Furthermore the number of intrusions into Japanese air space remains high: 812 in 1989 compared with 783 in 1980.[22]

The Northeast Asian security environment also embraces a segregated and volatile Korea. A leaked American Defense Department planning document recently identified the Korean divide as one of the major military challenges facing the United States in the future. Within days *The Economist* also described the Peninsula as perhaps the most dangerous place in the world.[23] But more revealing than the content of these statements was their timing. Both were made in late February 1992 after: North

Korea had accepted (September 1991) the arrangement of separate seats at the UN for the two Koreas; had signed an agreement (December 1991) to make the Peninsula free from nuclear
weapons followed by a safeguards agreement (January 1992)
with the International Atomic Energy Agency; and finally had
signed a reconciliation agreement (mid February 1992) intended
to be the first step to the Peninsula's reunification.

Notwithstanding these positive developments the potential for
conflict may now paradoxically be greater with the thawing of
the East–West permafrost. Germany's reunification could serve
to fortify North Korea, overriding the confidence-building measures introduced over recent times. In addition, economic, political and military pressures continue to build. There is the
widening north–south economic gap (North Korean per capita
GNP currently stands at 20 per cent that of South Korea's). This
may eventually lead to civil unrest in the communist north and a
precipitative military reaction from the south. On the political
front is the instability surrounding the inevitable succession of
the 'Great Leader', Kim Il Sung. Moreover, the fact that two-
thirds of Pyongyang's one million ground forces are deployed
near the Military Demarcation Line only a short distance from
Seoul heightens tension. In this respect, the United States' East
Asian Strategic Initiative, the planned diminution of US forces
in South Korea, provides the wrong signals to North Korea.
Finally there is the alleged North Korean development of nuclear weapons near Yongbyon. Although all these factors acting
individually or in unison destabilise, and invite misjudgement, it
is undoubtedly the nuclear weapons issue that is the most critical
in this regard.

North Korea signed the Nuclear Non-Proliferation Treaty in
1985, but refused to accept the mandatory safeguards that go
with it. In late 1991 after months of discussion Pyongyang
relented on its position, although only after protracted negotiations concerning the presence of US nuclear weapons in South
Korea. Revelations in September forced the North onto the
defensive. Information from a defector confirmed Western intelligence fears that the North is actively working on what has
been termed a 'decisive weapon', and is constructing a major
uranium reprocessing facility to extract plutonium at Yongbyon.
The South Korean Defense Ministry responded by issuing its
second warning in six months that 'at the worst, military action
may be taken' against North Korea's nuclear facilities if it re-

fused to halt its nuclear weapons programme.[24] North Korea has never admitted to a nuclear weapons development programme, calling into question the credibility of agreements between the Koreas to ban nuclear weapons from the Peninsula. As before, then, the fear of conflict continues, carrying the danger of drawing in other regional powers. Japan, in particular, watches the unfolding Korean drama with quiet alarm. Historically there has been enmity between the two nations, and this continues today. Japan has long exhibited cultural superiority to the Koreans. For their part, the Koreans, as public opinion polls invariably show, see Japan as one of their least liked countries, second in fact only to North Korea.[25] Moreover, at the official level, while the possibility of even peaceful reunification rekindles Japanese apprehension about Korea,[26] South Korea, by contrast, is concerned at what it perceives as the growing threat of Japanese militarism. This is evidenced by the 1991 South Korean Defense White Paper describing the JMSDF build-up as 'expansionist'.

Security tensions are not confined to Northeast Asia, however. Military developments across the Asia-Pacific region have created an uncertain strategic environment. There has been the withdrawal of former Soviet naval units from Vietnam's Cam Ranh Bay; also phased US troop reductions (triggered by the Nunn-Warner Amendments to the 1990 US Defense Appropriations Act) from South Korea, Japan (5000 to 6000 troops) and the Philippines (including here also the closure of Subic naval base and Clark air field). These events contribute to the evolvement of an Asia-Pacific security vacuum. Local military build-ups have been the consequence. For example: South Korea is procuring 120 F-16 fighters, around 100 Sikorsky helicopters, 16–18 destroyers, six HDW 209 class submarines and around 700 Type 88 MBTs; Taiwan is acquiring 260 aircraft, eight modified FFG-7 class frigates and around 400 MBTs; Thailand, F-16 fighters, hundreds of MBTs, early warning aircraft and Chinese frigates; Malaysia, 28 British Hawk jet fighters, with submarines also planned; Indonesia, up to 24 frigates, four submarines and fighter aircraft; and China, the purchase of a squadron of 24 Sukhoi Su-27 fighters from the former Soviet Union – the first time such advanced military technology has been made available to China by the former communist state, substantially upgrading China's power projection capability. It is thus difficult to envisage how substantial regional military expansion will allow Japan to reduce its military profile.

Figure 1.1 Japan's strategic environment

The existence of potential flashpoints and tensions do little to ease uncertainty. For example, less threatening though no less ominous is Brunei's procurement of 16 Hawk fighters ostensibly to protect offshore oil installations. The fighters are also required for territorial protection in the event of conflict escalation emerging from various disputed claims over the Spratly Islands (see Figure 1.1). Significantly, Brunei has now for the first time also laid claim to a portion of the Islands. Located some 200 km off the Bruneian coast, they are believed to contain oil and gas deposits. Brunei joins China, Vietnam, Taiwan, Malaysia and the Philippines in claiming sovereignty of some part of the Spratly Islands. Conflict looks increasingly possible. Indeed in 1988 Vietnamese forces did skirmish with those of China. Exacerbating tensions in this area these same two countries are also in dispute over the Paracel Islands in the Gulf of Tonkin. These are controlled by China, but claimed by Vietnam. China additionally, of course, lays claim to Taiwan, and also to the nearby Senkaku Island group, where again oil deposits are held to be located beneath surrounding seas. These Islands are also claimed by Taiwan and Japan. In pursuing these claims China would probably need an aircraft carrier. It is notable therefore that China has expressed an interest in the former Soviet carrier,

Varyag, continuing to be built even though surplus to requirements at a Ukrainian shipyard. Significantly, the SU-27 fighter can easily be operated from a carrier. As hinted at earlier, it is unlikely that Japan would not similarly respond in the face of serious escalation of regional naval capability.

Moreover, in the strategic shadow of these frictions is the burgeoning military power of India. Aggressive in its pursuance of regional hegemony the Indian navy is amassing a powerful blue-water capability. India possesses two aircraft carriers with a more advanced carrier planned for the mid-1990s. These carriers are equipped with Sea Harrier fighters and Westland Sea King helicopters. There are also over a dozen guided-missile armed cruisers and destroyers. The carrier task groups have support and replenishment facilities. India also has up to 20 diesel electric submarines and a nuclear-powered submarine. In addition, there are ocean-going assault tank landing ships, mine sweepers/counter-measure vessels as well as hydrographic research ships. Of regional concern is that the growth in India's naval power has occurred even though all its wars since Independence have been land-based conflicts. India's intention appears to be the extension of its maritime sphere of influence. This is to the consternation of China whose geo-strategic interests often conflict with India's. Evidence of this occurred in 1985 when a group of Chinese naval vessels visited Sri Lanka, Bangladesh and Pakistan signalling to New Delhi that the Indian Ocean was not exclusively India's Ocean.

The lack of stability spanning from the Koreas to the Subcontinent obviously affects Japan's security interests. The majority of Japan's trade is seaborne. Its vital oil and material inputs traverse the region's sea routes. Clearly, international trade represents the lifeblood of the economy. In an uncertain strategic environment Japan's trading dependence may oblige it to raise its military projection, but as yet there is not an inevitability about this course of action.

IV JAPAN's 'SUNRISE' MILITARY-INDUSTRIAL COMPLEX

Japan has risen phoenix-like above its Second World War industrial ashes to become one of the world's most powerful economic nations. Into the 1990s the global competitive war is almost won. The seedcorn for this particular campaign was the 'Yoshida Doctrine', with its thrust towards economic rehabilitation. Named after Japan's first postwar prime minister the doctrine advocates that economic revitalisation should be the priority goal, with long-term security dependent upon both the US nuclear shield and the presence of US forces in Japan. The question of whether minimal defence outlays assisted Japanese economic growth remains a controversial and unresolved issue. There is no doubting, though, that this factor along with others of an economic, social and cultural nature represented the appropriate ingredients for an industrial and technological miracle.

Japan's real national income in 1946 was 57 per cent of that achieved in the mid-1930s and early postwar foreign trade only 10 to 20 per cent of prewar levels; yet after allowing for the fact that industrial production had been decimated, coal output cut in half, and infrastructure and shipping almost totally destroyed, Japan had by 1954 already exceeded its highest prewar level of production.[27] The transformation of Japan's economy has proved remarkable, particularly in terms of the speed in which it has been effected. Japan's strategic approach has been an important contributory factor in this regard. Described by Chalmers Johnson as a 'capitalist developmental' model,[28] it adopts a subtle visible-hand approach, supplementing market forces rather than supplanting them, as under socialism. Technological infusion was promoted through licensing, while direct foreign investment was discouraged. Japan insisted on maintaining control over its technological destiny. The capital source for Japanese investment during the early decades after the war was capital that had been accumulated through suppressing consumption in favour of savings. Couple this with a strong work ethic, a pluralist production approach and supportive financial and institutional mechanisms and the conditions existed for rapid economic growth. The results have indeed been dramatic.

By the early 1990s Japan possessed after the United States the world's biggest economy. Even though its GDP was 60 per cent that of the United States' the gap is closing, and closing fast. In 1989 for the first time Japanese capital investment exceeded that of American. This means the catching-up process will accelerate. The United States' average annual growth rate of GDP between 1980 and 1990 was only 2.7 per cent, but Japan's over the same period was 4.5 per cent. As a consequence, while the United States' GDP share amongst the G6 countries (the United States, Japan, Germany, France, Italy and the UK) was 43.2 per cent in 1980 declining to 41.2 per cent in 1990, Japan's share over the same period rose from 17 per cent to 22.5 per cent, almost entirely at the expense of the United States.[29] Japan as No. 1 suddenly does not seem such an extravagant claim; achievable perhaps not far beyond the year 2000. Given that history illustrates the correspondence between economic power and military strength the question of Japan's military renaissance is thrust to the fore. Inevitably Japan's ability in this regard turns on the potency of its military-industrial complex.

Defence industrialisation is an important manifestation of Japan's military resurgence. Without even trying, its defence industrial base is among the world's five biggest. Strategically the justification for domestic arms production is security of supply production, and Japan has certainly achieved this. Local manufacture accounts for 99 per cent of naval ships, 89 per cent of aircraft and 87 per cent of ammunition.[30]

There are other reasons, however, for Japan's defense-industrial push. Amongst these economic ones predominate. The Japanese 'model', as coined in A. Edgar and D. Haglund's chapter (Chapter 8), is novel, containing several interrelated strands:

(1) As with civil sector development, defence development has been pursued through the mechanism of licensed production. This has principally been through United States–Japan collaborative effort. But it is an expensive option, given that the alternative approach of 'off-the-shelf' purchase would considerably reduce costs. The unit price of the new Type 90 MBT, for example, is $8.7 mn: 3.5 times more expensive than the export price of the M1-A1 Abrams.[31] Similarly the US–Japan agreement to produce the American F-15 fighter under licence has been estimated to have entailed costs per plane that were at least double those of direct purchase from McDonnell Douglas.[32] The Japanese regard this additional cost as a premium to be paid in evolving a dynamic comparative advantage.

(2) An important element of Japan's defence-development model is the technological synergy between the civil and military sectors. The Japanese were early exponents of technological 'spin-on', innovation flowing from civil to military application.

This approach, now much in emphasis by NATO and former Warsaw Pact defence industries, reverses the traditional view of defence spin-offs benefiting civil technologies; a thesis in any case of dubious practical value. The spin-on thesis, by contrast, implies that civil industrial capacity and technological infrastructure is already *in situ*, providing a ready source of cross-over resources for defence production.

There are two points to note here. Firstly, the 'strategic' civil industries servicing defence production have traditionally been regarded as the heavy engineering sectors such as iron and steel, non-ferrous metals and machinery production. Although these sectors are clearly still of great relevance, in the 1990s it is the dual-use industries such as micro electronics, ceramics, aerospace, computerisation and telecommunications that determine defence-industrial competence. Secondly, unlike Western defence contractors which have specialist defence manufacturing divisions the Japanese way has been to strive for integration, where both civil and defence programmes are pursued under the same roof with each activity cross-subsidising the other.[33] Under this approach technological spin-on becomes relevant at both the product and process level. Yet significantly the United States' superiority in these areas is fast eroding. For instance, a 1990 US Department of Defense study showed that at least 15 of 20 critical technologies surveyed related to dual-use items such as composite materials, semiconductors, and biotechnology processes, and of these Japan was significantly ahead in 25 per cent of the cases and capable of making major contributions in a further 15 per cent.[34]

Exploiting dual-use industries within the civil-military industrial penumbra promotes both lateral and vertical industrial linkages. Civil production continues to be emphasised but defence activities add to capacity utilisation, especially when the government pump-primes the economy in the event that civil demand slackens. Defence production also provides additional profitable opportunities particularly in an internal defence market that is oligopolistic in structure. The top Japanese defence five contractors account for over 50 per cent of total contracts and the top 10 garner 65 per cent.[35] Heading this list is Mitsubishi Heavy Industries accounting for 26.1 per cent of Japan's total defence contract value.

Japan's defence-industrial structure has been a technological attribute of some import. Stemming from the number and diversity of foreign collaborative contracts necessarily undertaken by the few Japanese prime contractors, the assimilation and integration of technologies transferred promotes their diffusion into the wider subcontracting base. But despite the relatively oligopolistic nature of the market, there are nevertheless well over a thousand suppliers (providing everything from aircraft, rice and shoes) to the JDA. The size of the subcontracting base has until only recently been growing, driven by the attractiveness of defence contracts, e.g. secure orders and assured cash flow. In addition, of course, there are technological benefits. Importantly, because this is a civil–military relation the benefits of technological progress have the potential of filtering down a further conduit into the civil manufacturing sector.

Moreover, Japan's mature and highly vertically disintegrated industrial structure is conducive to arms production requiring a diversified networking of specialist subcontractors working on refined engineering tasks to exacting quality specifications. In keeping with the technological dualism characteristic of Japan's civil manufacturing sector, defence final goods producers similarly exploit scale effects through dedicated transfer lines served by myriad small contracting firms utilising flexible manufacturing systems. The percentage of subcontracted work in Japanese programmes can be more than three-quarters of the total project value. This industrial pattern is highly efficient. It aims for both scale (prime contractors) and productive flexibility (subcontractors) – the 'magic kingdom' – mirroring Japan's successful civil industrial model.

(3) Yet another attribute of Japan's defence-development strategy is its thrust towards advanced research and development (R&D) acting as a catalyst for leading-edge technologies. Figure 1.2 shows the trend of Japanese defence R&D since 1976. Clearly, in symmetry with its rising defence budget, the JDA has devoted increasing amounts to defence in this area. Recently, R&D expansion has been identified as a major objective by the JDA; indeed it has been pressing for a near-doubling of R&D funds devoted to the Technical Research and Development Institute, JDA's specialist R&D branch.

Not unconnected, note that Japan's civil R&D spend is substantial; in fact comparable to that of the United States.[36] Indirectly, in a fused civil–military industrial infrastructure,

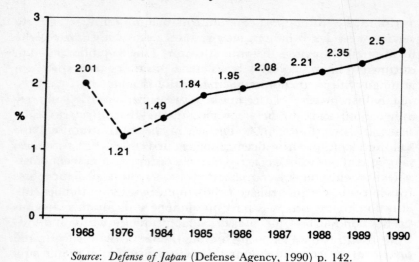

Source: *Defense of Japan* (Defense Agency, 1990) p. 142.

Figure 1.2 Changes in terms of the ratio of the budget of the Technical R&D Institute to defence-related expenditures over the years 1968–90

Japan's civil R&D expenditure benefits defence technology. This is because much of the underpinning dual-use technologies have had their development leveraged by high levels of civil R&D funding.[37] The electronics sector provides a good example in this respect. The fruits of the sector's rapid technological development over the past two decades have not been isolated to civil production. It has been estimated, for example, that the share of electronics in total weapon cost is 23 per cent for the T-74 tank, 21 per cent for the F-4 aircraft and 40 per cent for the P-3c aircraft.[38] Albeit that a proportion of these electronic technologies would have been transferred from the United States, a good measure would also have come from local capacity. This is supported by the fact that 80 per cent of the FSX including all electronics will comprise indigenous technology.[39]

During the 1980s Japanese defence industrialisation has been gradual, but continuous. Defence output as a share of total output grew from below 0.4 per cent to almost 0.6 per cent between 1980 and 1987. A small relative proportion, yes, but as was mentioned earlier, in absolute terms Japanese defence expenditure could soon become the world's second biggest: 1 per cent of Japan's GNP is a huge amount. In 1990 Japan followed

India as the leading importer of armaments. Yet after actively encouraging Japanese rearmament, such statistics unnerve the US Congress: Japan's 'no-win' dilemma. This has contributed to strained relations, with US reactions spanning the spectrum from envy to racist innuendo and finally to fear.

US insecurities emerge most dramatically in the military sphere, but more likely are rooted in the economic area. The fears are that US relative decline in the civil sector will be followed by a Japanese defence-development push. The scenario is pregnant with *déjà vu*. Stage one: the absorption of technological knowledge developed elsewhere, with the transfer process driven primarily by licensed production. Stage two: import substitution giving way to export promotion, with Japan's virtuous production-marketing cycle taking over – high-quality and low-priced output leading to increasing demand, scale effects, still lower cost-prices and eventually higher demand; Japan's proprietary knowledge in the process being ring-fenced.

The concern, increasingly, is not whether Japan will adopt this strategy in defence markets, but when. However, a more measured perspective is perhaps appropriate. Policy developments will be dependent upon: domestic and international reaction to Japan relaxing its restrictive arms export policy; the achievement of increased security and diplomatic profiles; and the general status of US–Japan relations.

Article II of the 1954 Mutual Security Treaty states that an objective of the Treaty was mutual economic advancement. Has this been achieved? Although this is a contentious issue, pointers exist to suggest that the Security Treaty has been of enormous mutual value. It is a widely disregarded fact, for instance, that Japan represents the United States' biggest overseas export market, second only to Canada in overall purchase of US products, both 'high tech' and 'low tech'. Unquestionably, also, the strategic stability the Treaty provided has enabled a broader and more sustained degree of economic growth throughout the Asia-Pacific region. This argument applies *a fortiori* to Japan. In 1991 Japan's trading surplus was $78.2 billion, almost 50 per cent of which was earned in the US market.[40] Indeed, it has been Japan's economic prowess that has unfortunately earned it the ire of the US (and Europe alike). The placing of Japan on the US Super 301 list, naming it as an unfair trading partner and banning the imports of certain of its products into the United

States was serious, but the FSX *débâcle* probably carries greater long-term implications. Whether or not the FSX advances beyond the prototype stage, the acrimonious political after-taste will linger on, certainly lessening the possibilities of similar major United States–Japan defence collaboration taking place in the future. The impact on Japan's defence procurement policy will be two-fold. Firstly, a more 'involutionary' approach will be pursued. In this regard note the 1991 JDA announcement that the JGSDF's light observation helicopter will be indigenously designed and manufactured, and that in 1992 basic research was begun by the Technical Research and Development Institute of a locally developed fighter with superior manoeuvrability and stealth characteristics to the USAF's Lockheed YF-22 Advanced Tactical Fighter. Secondly, Japan may plump more aggressively for diversification of defence-technology sources.

Years of US technology transfer have led to the reality of Japan's defence-development 'take-off'. But in consonance with its one-sided interpretation of obligations under the United States–Japan Defense Treaty (aversion to an 'Alliance'; that is, US military forces aiding Japan if attacked, but without reciprocity in the event the US were attacked), technology flows have also been less than equal. The fact is that the 1983 exchange of notes on military technology transfers between Japan and the United States, whereby a two-way transfer of defence technologies was permitted, has in practice facilitated few transfers from the Japanese end. Possibilities for raising the value and pace of Japan's technology flows are bedevilled by policy difficulties including a narrow interpretation of military technologies as well as an official embargo on third country resales. It is difficult therefore to take a sanguine view of future United States–Japan defence cooperation. Indeed bilateral difficulties may increase as Japan's defence-industrial capability and international activism develops. The United States will then cease to be Japan's collaborator, becoming more explicitly its competitor.

V INTERNATIONAL CONTEXT

The final aspect of Japan's comprehensive security approach relates to the political-diplomatic sphere. If Japan, in a multipolar world of regional 'Great Powers', is to assume a role commensurate with its economic power then political normalisation is required. This means becoming an 'equal' in the international political and security systems. In fact if the 'vacuum' theory is credible then in line with the United States' relative economic decline and strategic withdrawal will come a heightened Japanese regional political and security profile; the process auguring a gradual enlargement of Japan's defence capabilities in harmony with its expanding international responsibilities. This view suggests that Japan has outgrown the 'Fukuda Doctrine'. The Doctrine espoused by Prime Minister Fukuda in 1978 represented an attempt for the first time to develop a policy framework for relations between Japan and its Asia-Pacific neighbours. It aimed to promote political and diplomatic activity in the region, falling short however of a greater military presence. In the 1990s the Doctrine looks decidedly anachronistic: geo-strategic changes; Japan's growing regional influence; and its increasing emphasis on economic security, encompassing 'food security'[41] have led to calls for a broader remit for the JSDF. These demands surfaced during the early 1980s when Prime Minister Nakasone initiated the challenge to Japan's postwar political insularity by pressing for normalisation. Nakasone's successor, Noboru Takeshita, consolidated this position by advocating internationalism, embodied in his 1988 'International Cooperation Initiative'. The Initiative centred around the provision of international assistance, which in the context of Japan has been described by one observer as 'strategic aid'.[42]

The Initiative has three pillars: expansion of Official Development Assistance (ODA), promotion of cultural exchange, and strengthening of Japan's contribution to international peace.[43]

In the years following this policy announcement much has been done to achieve these aims. Japan's ODA has expanded dramatically: six times in terms of Yen between 1977 and 1988.[44] The third medium-term plan was to double ODA between 1986 and 1992 to more than $40 bn. In the event this was achieved by 1988. The target then became $50 bn by 1992. Eventually the aim is to achieve an ODA share in total Development Assistance Committee (DAC) countries' ODA corresponding to the share of Japan's GNP in that of the DAC members. Since 1989 Japan has become the world's biggest ODA donor dispersing around $10 bn annually. In 1988, 54 per cent of Japan's ODA went to Asia, representing for that region more than the rest of the world combined.[45] At the cultural level funds have been dispersed worldwide to foster links such as Japanese Exchange and Teaching (JET) programmes at European and North American universities and preservation projects under the aegis of UNESCO. Although it is less easy to identify Japan's specific contributions to securing international peace the country's $13 bn contribution to support the Gulf War clearly comes under that category. Indirectly, also, Japan's participation in a multilateral aid package to the Philippines must be included. Moreover, there is no question that a political momentum is building up for Japan to exert greater influence in the resolution of regional conflicts. In this context, JSDF involvement in UN-sponsored multinational peacekeeping forces, monitoring support and observation teams, appears likely sometime in the future.

A more regional security focus seems inevitable. Part of the reason for this is economics. The Yen's appreciation coupled with rising domestic labour costs during the 1980s have led to huge amounts of Japanese foreign direct investment (FDI). In 1991 Japan was the world's biggest holder of net foreign assets, amounting to $383.1 bn, a position it had held for six of the previous seven years.[46] ASEAN countries in particular have attracted substantial capital flows. For instance, between 1984 and 1989 Japan's FDI in Indonesia amounted to $9.8 bn, approximately a third of Japanese FDI in all the countries of Europe.[47] Similarly for Singapore: in 1988 FDI there accounted for 83 per cent of total investment with Japan's share being almost 35 per cent.[48] This and the substantial trading surplus Japan enjoys with its Asian neighbours is creating a Yen-denominated Asian trading area of some importance. Ironically, then, economic power is in the throes of achieving Japan's prewar military goal: a 'Greater East Asia Co Prosperity Trading Sphere'.

Dependence and the increasing density of regional economic tentacles perhaps justifies Japan's concern over market and supply security. Existing security arrangements look fragile. The longstanding Five-Power Agreement is more a parochial anachronism than an all-embracing Alliance, while ANZUS (Australia, New Zealand and the United States) has suffered a credibility crisis since the political disagreements between New Zealand and the United States in the 1980s. There is also, of course, the declining US military influence. This was highlighted in 1991 when the US Defense Secretary, Mr Cheney, announced plans to withdraw 10 per cent of US forces from Asia. But this only perpetuates a trend begun in 1967 with the enunciation of what became known as the 'Nixon Doctrine'. At that time President Nixon emphasised that the United States, while reaffirming its treaty commitments in the region, insisted that, in future, Asian countries would be expected to assume the primary responsibility for providing the manpower for their defence. In the 1970s there followed three further 'Nixon shocks': the President's plans to visit China; surcharges on American imports and the suspension of dollar convertibility to gold; and a US ultimatum during textile trading negotiations threatening Japan with application of the Trading with the Enemy Act. These events along with the more contemporary bitter FSX debate and incessant 'Japan-bashing' has created much unpredictability in United States–Japan relations. Combine this with the overstretching of US international financial obligations, and a reducing American presence in the Asia-Pacific region should come as no surprise.

The perceived security void although not yet a reality is no less problematical. While in the long term Japan aims to obtain a permanent seat on the UN Security Council, in the short run it has taken the first tentative steps towards promoting a regional security consensus. At the ASEAN meeting in July 1991, Japan's Foreign Minister suggested the establishment of a 'security forum' for Asia-Pacific countries. The Japanese proposal was based on the argument that Asia needs a security mechanism akin to the Conference on Security and Cooperation in Europe (CSCE). Yet the differences between Europe and Asia are immense. In Europe, spatial, cultural, economic and security factors converge. In the South, by contrast, ASEAN countries as well as other states speckle oceans spanning 180 million square kilometres; their religious, ethnic, cultural and industrial characteristics differ considerably; and the type of neutral and non-aligned (N + N) countries which played an important role as mediators in the CSCE process do not exist among the Asian-Pacific nations.[49] Not surprisingly, therefore, ASEAN's reaction to Japan's proposal was one of studied ambivalence, but security did get on the agenda and will likely return at future gatherings.

Memories endure. The former Japanese colonies in Asia will not lightly forget the political trauma and human suffering now stigmatising Japan's Armed Forces. Asia's uncertain geo-strategic environment and the realities of hard-nosed economics ensure a greater regional role for Japan. Events such as goodwill calls at Asean ports by JMSDF minesweepers whilst on their way to the Gulf; offers of financial and physical support for UN monitoring of the international arms trade; mediation in resolving the Cambodian problem; and also involvement in Korea's 1991–2 normalisation talks represent Japan's beginnings in establishing a new era in regional perceptions and relations.

Finally, the Japanese quest for normalisation must overcome widespread prejudices, misconceptions and misunderstandings. Japan's political and security contradictions compound the enigma. Yet international recognition of the changed political and economic circumstances is paramount. For potentially the world's most powerful economic nation to continue into the next century in a state of military emasculation is difficult to rationalise, save for the continued martyring of guilt borne half a century earlier.

IV   STUDY AIM AND OUTLINE

This introductory *tour d'horizon* of the subject-matter has attempted to lay the foundation for the detailed chapter discussions which follow. Although the substance of individual chapters focuses on only one particular strand of the debate, clearly all chapter topics interweave into the overall fabric. A macro, strategic perspective is crucial to understanding the overlay of considerations. The notion of military security is perhaps too confining for the international environment of the 1990s. For Japan, the military comprises only one of the variables in the security equation. Others embrace politics, economics and international diplomacy, widely defined. Of course, traditionally, security has always touched on these related areas. What is so different with the Japanese approach is the relatively low weight given to the military dimension.

Naturally this could change. Japan is very much at the crossroads. Astute and flexible international diplomacy may lead to successful outcomes of such thorny issues as Korean reunification and CIS–Japan relations. In reference to the latter, with skilful negotiation Japan might even be able to secure the return of the Kurile Islands as a *quid pro quo* for foreign aid and investment in 'kickstarting' the former Soviet economy. On this basis, then, the disputed Spratly, Paracel and Senkaku Islands could replace the Northern Territories as the central strategic issue shaping Japan's future policy direction. But this is not to diminish the country's other foreign policy imperatives. Japan's prickly relations with the United States, the status of the US–Japan Security Treaty and the robustness of the New World Order are also factors helping to forge Japan's evolving military profile. Can this profile be minimised, however?

Will Japan be unique in world history: as a major economic power can it break the conventional mould, continuing into the twenty-first century with a benign, insular foreign policy framework? This is a difficult question to answer, especially given the current turbulent international environment and the dynamic nature of Japanese society and institutional processes. But the question is important as there are policy-issues embroiled here which cannot be avoided. The aim of this book is to address them.

The book's contributors represent some of the world's leading Japanologists in the security arena. They write authoritatively and dispassionately, offering perceptions from four continents. The problems faced by Japan in successfully integrating into the regional and even global security network are not minimised. Japanese integration will be difficult, but also perhaps inevitable. More than anyone, the Japanese themselves need to recognise this.

It is for this reason that Japanese authors were invited to participate in this book, to make their own contribution to the debate. Professor Watanabe's chapter represents the first of several Japanese perspectives (Chapter 2). He deals directly with the constitutional issue, arguing that although the Japanese Constitution has reflected well the needs of the people, the time has now arrived for its reinterpretation. Japan's economic power necessitates enlargement of its international responsibilities, but through (international) collective security not collective self-defence. Ian Gow's contribution (Chapter 3) also focuses on the constitutional issue, but with emphasis on the important topic of civilian control of the military. By reference to Japan's role in the Gulf crisis, the author reaches a similar conclusion to the previous writer. To avoid undermining the civil–military control framework, a differing interpretation rather than constitutional change is the appropriate path to greater flexibility in security policy. Chapter 4 by Professor Sakanaka also advocates Japan adopting a more flexible response to international affairs, but for this to be effective Japan's role in the international division of labour needs to be formalised. The author argues that the present climate offers Japan a 'golden opportunity' for this to occur. Further examining Japan's security role, especially in relation to ASEAN, the Australian observer, Dr Maswood (Chapter 5), raises the possibility that the Japanese may be obliged to become a major military power to ensure conditions conducive to their economic interests. He tempers this judgement, however, by further stating that Japanese militarisation should be pursued within the United States–Japan security framework, itself being part of a wider multilateral system. An important by-product of this would be that a more equal, and more assertive, Japan could act as a counterweight to US exuberance. In Chapter 6 Dr Segal grapples with similar aspects of this theme, but from the standpoint of the New World Order. He also adds his voice to those who advocate that the time is ripe for changes in Japanese foreign policy; and this, of course, has political and security consequences. No longer, the author states, is it reasonable that Japan is 'merely expected to pay and not shape the World Order'.

The discussion then steers away from purely strategic security towards related defence-industrial and technology aspects. This changed focus begins with the chapter by Professors Matsuyama, Kojina and Fukuda (Chapter 7), tackling the controversial subject of the 'trade-off' between defence expenditure and economic growth. Interestingly, the authors' quantitative analysis suggests that since 1980 Japan's strong growth in its military budget has had no major impact on the increase of national income. At least a partial explanation for this may derive not simply from the level but also the nature of Japanese military-industrial activity. Professor Haglund and Alistair Edgar dwell on this theme in Chapter 8. They identify what they term a 'Japanese model' of defence industrialisation, the thrust of which is directed towards the civil linkages of the military-industrial relation. Moreover, the authors tellingly reveal that it is MITI, the Ministry of Munitions during the Second World War and the orchestrator of Japan's successful postwar civil economic strategy, that is 'pulling the strings' of Japanese defence-industrial development. Similar success in the defence field holds the potential for exacerbating competitive tensions with, in particular, the United States. Professor Hartley and Dr Stephen Martin's chapter (Chapter 9) is one of a number that express the view that US–Japanese friction over technology transfer will probably precipitate the growth in Euro-Japanese defence collaboration. These authors focus their study on the aerospace sector, an area in which Japan has admitted to ambitions. Michael Chinworth in Chapter 10 examines in detail US–Japan Technology transfer problems in the defence collaborative field. The United States is no longer easily able to accommodate the economic risks associated with defence technology leakage from the United States to Japan. Previously the Soviet threat overshadowed these concerns, but no longer. The penultimate chapter falls to Professor Taylor, who examines Japan's domestic policy stance on military and military-related exports (Chapter 11). The author develops the discussion by arguing that Japan is in the vanguard of international attempts to control the arms trade, but there are practical difficulties, such as those of implementation, the achievement of 'transparency' and the requirement that arms restraint applies equally to all nations – including all members of the UN Security Council. Significantly, Professor Taylor also notes that as Japanese defence-industrial expertise develops apace with its political assertiveness then the spectre looms of Japan vetoing US arms sales (incorporating a high percentage value of Japanese components) to third countries; a 'heady' but entirely possible prospect.

Concluding the book is a contribution by Professor Endicott, who writes on the changing social and cultural characteristics of the Japanese people. Implicitly, he questions whether Japanese economic growth may have 'peaked' through the emergence of individualism, a fading work ethic and a less focused corporate commitment: the *shinjinrui* or 'new breed' of young Japanese worker rejecting the role model of the tireless but now ageing salaryman. Professor Endicott speculates on the influence of these factors in shaping Japan's international relations and foreign policy. In particular, he highlights the dangers of the eroding American and Japanese public perceptions of each other. Mutual trust is disintegrating at the people level. US 'Japan-bashing' is being met by the rebirth of the Japanese 'superiority-syndrome'. Both trends are disturbing.

Successful resolution of these and other problems discussed in this book, rests ultimately on the acceptance by the Japanese *and* the international community of Japan's role in the New World Order. In this context, then, the generalised consensus of the various author arguments presented here is that the dawning of a new era is nigh, but this time around the 'rising sun' will exorcise the ghost of Japanese militarism.

Notes and References

1. I am indebted to Michael Chinworth and Col. Stewart Hendry for their comments on earlier drafts.
2. Stefan Wagstyl, '"Illiterate and Lazy" Jibe Stokes Tension with US', *The Financial Times*, 22 January 1992.
3. S. Javed Maswood, *Japanese Defence – The Search for Political Power*, Institute of Southeast Asian Studies (Allen & Unwin, 1990) p. 79.
4. See Hideo Sato, 'Japan's Role in the Post-Cold War World', *Current History*, vol. 90, no. 555, April 1991, p. 146. Original source: David P. Rapkin, 'Japan and World Leadership?', in D. P. Rapkin (ed.), *World Leadership and Hegemony*, International Political Economy Yearbook, vol. 5 (Boulder, Col. and London: Lynne Rienner Publishers, 1990) pp. 196–9.
5. James E. Auer, 'Article Nine of Japan's Constitution: From Renunciation of Armed Force "Forever" to the Third Largest Defence Budget in the World', *Law and Contemporary Problems*, vol. 53, no. 2, Spring 1990, p. 172 fn.
6. S. Javed Maswood, *Japanese Defence*, p. 8. Original source: H. Otake, *Nihon no Boei to Kokunai Seiji: Detente Gunkakue* (Japanese Defence and Domestic Politics: from Detente to Militarization) (Tokyo: Sanichi Shobo, 1983) p. 39.
7. James E. Auer, 'Article 9 . . .', p. 177.
8. Ibid, p. 177.
9. S. Javed Maswood, *Japanese Defence*, p. 61.
10. Ibid, p. 70.
11. See *The Financial Times*, 7 December 1988, and the *Far Eastern Economic Review*, 13 October 1988.
12. 'The Cautious Sharpening of the Samurai's Sword', *The Economist*, 16 August 1986.
13. Ian Rodger, 'Japan's Military Chips Under Fire', *The Financial Times*, 29 January 1991.
14. G. Counsell, 'Japan to Cut Dependency on Energy Imports', *The Independent*, 25 April 1991.
15. James E. Auer, 'Article 9 . . .', p. 178.
16. US Congress, Office of Technology Assessment (OTA), *Arming Our Allies: Cooperation and Competition in Defense Technology* (Washington D.C.: US Government Printing Office, May 1990) p. 108.

17. J. T. Bergner, *The New Super Powers – Germany, Japan and the US and the New World Order* (St. Martin's Press, 1991) p. 174.
18. James E. Auer, 'Japan's Defense Policy', *Current History*, vol. 87, no. 528, April 1988, p. 147.
19. Ibid, p. 148.
20. It is interesting to note that Japan's territorial interest has since 1931 always lain further south than the offshore Honshu. Okinotorishima, a sprinkling of tiny rocks 1000 miles south of Tokyo, justifies Japan's claim to 160 000 square miles of ocean. The importance attached to these islands was illustrated in 1988 when the Ministry of Construction spent around $200 mn encasing the rocks in concrete to stop them sinking beneath sea level. Robert Whymant, 'Japan Crowns Isle', *Daily Telegraph*, 28 May 1988.
21. Dr Young Koo Cha, 'The Changing Security Climate in Northeast Asia', *International Defense Review*, 6/1991, p. 616.
22. *Defense of Japan 1990*, Government White Paper, Defense Agency, 1990, p. 132.
23. 'A Hesitant Patroller of the Pacific', *The Economist*, 27 July 1991, pp. 45–6.
24. Joseph Bermudez, 'N. Korea On Way To "Decisive" Weapon', *Jane's Defence Weekly*, 12 October 1991, p. 653.
25. C. Kim, 'Korea–Japan Relations and Japan's Security Role', *Korean and World Affairs*, vol. 12, Spring 1988, pp. 105–6.
26. Complete withdrawal of US troops from Japan is unlikely to occur in the foreseeable future. Budgetary reasons just as much as strategic interests account for this: Japan now meets almost 50 per cent of US costs through the 'Host Nation Support Programme'; it would be more expensive to re-station US forces in the Continental USA.
27. J. T. Bergner, *The New Super Powers*, p. 111.
28. Chalmers Johnson, 'The People Who Invented the Mechanical Nightingale', *DAEDALUS*, vol. 119, no. 3, Summer 1990, p. 74.
29. 'Rise and Fall of the World's Economic Superpowers', *The Financial Times*, 4 November 1991.
30. See Reinhard Drifte, *Arms Production in Japan: The Military Applications of Civil Technology* (Boulder, Col.: Westview Press, 1986).
31. 'Moving Against the Flow: Japan's Defence Build-Up', *Jane's Defence Weekly*, 10 August 1991, p. 232.
32. B. Udis and K. Maskus, 'Offsets As Industrial Policy: Lessons From Aerospace', *Defence Economics*, vol. 2, no. 2, 1991, p. 160.
33. Note that in Mitsubishi Heavy Industry's Nagoya Works, F-15 production takes place alongside subcontract work for Boeing aircraft. The airframe of the Boeing 767, for example, is made with 15 per cent Japanese parts. Commercial Aerospace now accounts for around 30 per cent of Japan's aviation industry's activities. See 'A Yen for Arms', *Far Eastern Economic Review*, 22 February 1990, p. 59.
34. See US Department of Defense, *Critical Technologies Plan*, Committee on Armed Services, US Congress, 15 March 1990.
35. OTA, *Arming Our Allies*, p. 104.
36. 'Japanese Technology Survey', *The Economist*, 2 December 1989, p. 4.
37. 'Japanese Defense Industrial Policy and US–Japan Security Relations',

US Congress, Office of Technology Assessment, *Global Arms Trade* (Washington D.C.: US Government Printing Office, June 1991) pp. 107–20.

38. Gregory P. Corning, 'US–Japan Security Cooperation in the 1990s – The Promise of High-Tech Defense', *Asian Survey*, vol. 29, no. 3, 1989, p. 278.
39. Ibid. p. 281.
40. Robert Thomson, 'Japan's Trade Surplus Up 50 per cent', *The Financial Times*, 22 January 1992.
41. Although Japan is the world's biggest importer of agricultural products, to GATT's anguish, it still maintains a rice market heavily protected from overseas competition. See Kevin Rafferty and Larry Elliot, 'Japan Fuels GATT Acrimony Over Rice', *The Guardian*, 18 February 1992.
42. See, D. T. Yasutomo, *The Manner of Giving: Strategic Aid and Japanese Foreign Policy* (New York: Lexington, 1986).
43. Tsuneo Akaha, 'Japan's Security Policy After US Hegemony', *Millennium*, vol. 18, 1989, p. 441.
44. *Japan's ODA, Overview* (Japan Information Office, London, 1989).
45. Calculated from *Japan's ODA, Overview*, ibid.
46. Stefan Wagstyl, 'Japan Tops Foreign Assets League', *Financial Times*, 27 May 1992.
47. J. T. Bergner, *The New Super Powers*, pp. 138–9.
48. Ron Matthews, 'Singapore's "Indigenous" Technological Development', unpublished mimeo, National University of Singapore, 1990.
49. See B. Sanwerwei, 'A CSCE for Asia, *International Defense Review*, May 1991, p. 54.

2 Japan's Postwar Constitution and Its Implications for Defence Policy: A Fresh Interpretation

Akio Watanabe

I INTRODUCTION

'Sea change' seems an appropriate cliche to describe the present international scene. Concrete forms in which such a change manifests itself vary, however, from place to place and from issue to issue, making it hard to guess the eventual shape world affairs will take. With this generalised feel of global change as a background, one cannot fail to be struck by the remarkable continuity of one important aspect of Japanese body politic: the Constitution.

Since the introduction of the modern concept of constitutional government about a century ago, the Japanese have had two Constitutions: the Meiji Constitution promulgated on 11 February 1889, and the postwar Constitution issued on 3 May 1947. Granted by the Emperor Meiji, the former had for long been believed to be something 'sacred and inviolable'. No one dared even to discuss the possibility of its amendment. The Meiji Constitution had been so 'hard' that when a drastic change was needed under postwar circumstances no effective mechanism for amendment was easily available. The result was the 'imposition' by an external power, i.e. the Occupation authorities, of the 1947 Constitution. Unlike the 'sacred and inviolable' 1889 document, the postwar document has been a target of constant criticism primarily due to its dubious birth. The irony, however, is that the so-called MacArthur Constitution well catered to the needs of the Japanese public, showing (thus far) great resistance to any suggestion of constitutional amendment. But in any case, given

35

the provision that calls for a two-thirds majority of the Diet for initiating amendments before being rendered for ratification by popular vote (Article 96), it is practically impossible for government, unless backed by an extraordinarily strong government party, to amend the Constitution. Thus, to the postwar Japanese the Constitution is also 'sacred and inviolable', albeit in an entirely different sense from the earlier example.

For the reasons outlined above, the Japanese have a very 'hard' Constitution in comparison not only with the British-style 'soft' Constitution but also with the relatively 'hard' ones of Continental Europe. A question arises, therefore, whether and how the Japanese would be able to adapt their 'peace Constitution' with the changing realities of world affairs. There are no statistical data, but a general impression is that a majority of Japanese scholars in the field of constitutional law are of the opinion that adaptation is neither necessary nor desirable.[1]

Being a scholar interested in international affairs, the present writer has a somewhat different view. Official views as well as public opinion in Japan are apparently in a state of flux, presenting a confusing picture. What follows is therefore neither an attempt to introduce what seems to be a prevailing view among constitutional law scholars nor an effort to provide a generalised view, in a statistical sense, of the Japanese people at large. Instead, it is the author's personal interpretation of Japanese constitutional problems in connection with international security issues. The discussion is in two parts. In the section that immediately follows, the problems of Japan's peace Constitution under the strains of the Cold War are examined. Following that, discussion is directed at how the ending of the Cold War is affecting Japanese thinking on issues of international security and the peace Constitution.

II JAPANESE PEACE CONSTITUTION UNDER THE STRAINS OF THE COLD WAR

A contentious issue surrounding Japan's 1947 Constitution has regard to the origins and intent of Article 9, which reads *in toto*:

Aspiring sincerely to an international peace based on justice and order, the Japanese people forever renounce war as a sovereign right of the nation, and the threat or use of force as a means of settling international disputes.

In order to accomplish the aim of the preceding paragraph, land, sea, and air forces, as well as other war potential, will never be maintained. The right of belligerency of the state will not be recognized.

This should be compared with the corresponding clause of the draft constitution handed to the Japanese government by General Douglas MacArthur's GHQ on 13 February 1946, which read as follows:

War, as a sovereign right of the nation is abolished. The threat or use of force is forever renounced as a means of settling disputes with any other nation.

No army, navy, air force, or other war potential will ever be authorized and no right of belligerency will ever be conferred upon the State.

We are not concerned here with the details of the drafting process of the 1947 Constitution. Suffice to say that:

(a) the GHQ's draft was radically different from other drafts, including those earlier ones prepared by the Government Commission, which preceded it;
(b) the war renunciation clause (Article 9) as stated above is essentially identical to that of the draft constitution crafted by the GHQ;
(c) there are nevertheless modifications to the original text of Article 9, indicative of reservations felt by the Japanese lawmakers at the time.

The first two of the above-mentioned points are concerned with the origins of the war renunciation clause, whilst the significance of the last has regard to the intent of that provision. The immediate event that prompted the GHQ to prepare its own draft constitution was the leakage by a Japanese newspaper of a draft that was under consideration by a Japanese Government Commission (1 February 1946). Realising that the Japanese draft was nothing more than a modest amendment of the Meiji Constitution, General MacArthur decided to intervene directly in the matter. He ordered the GHQ's Government Section to prepare a draft based on the three principles he had jotted down on a notepad. These principles were concerned with the status of the Emperor, war renunciation and abolition of the hereditary privileges of nobility (3 February 1946). Only ten days after the newspaper leak the GHQ submitted its draft constitution, taking the Japanese government by surprise.

A central feature of MacArthur's draft constitution was the notion of war renunciation, for which there are two possible explanations. Firstly, it may well be that this had been a long-held view of the experienced general, who sought not to miss the chance of realising his conviction. A second possibility, as MacArthur himself recollected later, is that he became possessed by this idea of war renunciation during a conversation with Baron Kijuro Shidehara. This took place when the latter (Prime Minister of the day) paid a courtesy visit to express gratitude for a gift of penicillin given to aid his recovery from a bout of influenza. Historians have so far been unsuccessful in finding the original text of MacArthur's 3 February note, resulting in long-standing ambiguity and controversy surrounding the contents of this note. It is, however, a well-known fact that, anticipating the commencement and mood of the Far Eastern Commission (FEC) a few weeks later, MacArthur needed something dramatic in order to pre-empt hostile criticism by some members of the FEC about his policy of preserving the Emperor-system. This immediate political concern of the general was precisely the reason why he decided rather abruptly to intervene personally in Japan's constitutional reform. War renunciation was an ideal proposal that would have satisfied his purpose. In this sense, then, Article 9 of the Japanese Constitution appears more a product of historical accident than the outcome of considered deliberations and planning.[2]

However, historical accident or not, the peace Constitution could not have survived beyond the period of its immediate usefulness if it had been totally out-of-tune with the public and political sentiment of the times. At first glance it may appear that the Japanese Constitution's war renunciation clause was something extraordinary; its existence explained by political expediency: the need to ward off otherwise unavoidable criticism of pressures upon the Emperor-system. Without denying the importance of such extremes in determining the course of history at a particular moment, there exists a further influence predating these events by decades. This is the Kellogg-Briand Pact in which the signatories agreed to renounce war as an instrument of national policy and to settle all international disputes by peaceful means.

It is probable that men like General MacArthur (1880–1964) and Baron Shidehara (1872–1951), both of whom were actively engaged in international activities in the post First World War era, were familiar with the war renunciation pact which reflected the pacifist sentiments of the day. It should be remembered also that this 1928 treaty greatly influenced the judgement on crimes against peace of the International Tribunal that tried the German and Japanese war criminals at Nuremberg and Tokyo. (These tribunals were in progress when the Japanese Constitution was being discussed.) In addition to such circumstantial evidence, there is also direct evidence showing the influence of the Kellogg-Briand Pact on the war renunciation clause of the Japanese Constitution. It is known that Colonel Charles L. Kades, Harvard Law School graduate and an aide to General MacArthur, was responsible for drafting that particular clause of the Constitution. Kades recalled that

> [the Kellogg-Briand Pact] made a deep impression on me when it was signed in 1928. I was in law school then, and I thought 'this is going to be an era of peace' because of the Kellogg-Briand Pact.[3]

The reason Kades was recalling the important bearing the 1928 treaty had upon the 1947 Japanese Constitution should be made clear at this point. During a 1980s interview with Professor Nishi, Kades admitted that in the process of formulating the GHQ draft he reminded his colleagues of the Kellogg-Briand Pact. The Pact did not allow for aggressive war but did allow defensive forces. Impressed by this distinction, Kades accordingly made 'a few little changes' in MacArthur's original instruction. MacArthur's three-point note of 3 February had stated in relation to the war renunciation that 'Japan renounces it [war] as an instrumentality for settling its disputes and *even for preserving its own security*' (emphasis added). Kades omitted the words 'even for preserving its own security' and added 'threat of use of force', because it seemed to him it was not realistic to say that if Japan were attacked, it could not defend itself.[4]

Significantly, MacArthur did not overrule his subordinate's modifications which were in fact more than 'a few little changes'. A further significant amendment was made during the course of Diet deliberations. The Diet adopted a proposal by Hitoshi Ashida, ex-diplomat and Chairman of the Special Committee of the Imperial Diet, to investigate the Constitution Amendment Bill. Ashida proposed that the words 'Aspiring sincerely to an international peace based on justice and order, the Japanese people' should be inserted at the opening of the first paragraph, and the words 'For the above purpose' at the beginning of the second paragraph. By adding these words (particularly the second group), Ashida meant to indicate that the Constitution would not prohibit the use of armed forces for all purposes.[5] The intent of the Ashida Amendment did not pass unnoticed by the GHQ officials. Realising that it opened the way for Japan to maintain war potential for the purpose of defence, the GHQ decided it was necessary to 'put on the brakes', and ordered the Japanese government to insert the so-called civilian clause into the Constitution. This is the origin of Article 66, stipulating in part that 'the Prime Minister and other Ministers of State must be civilians'.[6]

To those who participated in the formulation of Article 9, it would have not have seemed unnatural for Japan, condemned by the international community for violation of the Kellogg-Briand Pact and the League Covenant, to write the spirit of those international treaties into its new Constitution.[7]

At the same time, however, practical men like Kades and Ashida were aware that as a matter of historical fact signatories to the war renunciation pact allowed themselves a variety of qualifications and interpretations so that it would neither prohibit wars of self-defence nor collective action under the League Covenant. Yet the Constitution's architects could not afford to ignore political expediency. The Japanese, and by extension their custodian (the Occupation authorities), were obliged to impress world opinion about the genuinely pacifist nature of the Japanese body politic under the new Constitution. This explains the existence of ambiguity and inconsistency in the wording of

Article 9. Clearly, to re-emphasise, it was a product of political compromise.

It is no wonder, therefore, that Shigeru Yoshida, the then Prime Minister, emphatically stated in response to an interpellator that it would be dangerous to make a distinction between wars of aggression and those of self-defence because almost all wars in recent history were fought in the name of self-defence.[8]

Japan's right of self-defence remained purely an academic question as long as the country was under the custody of American occupation forces. It would remain so forever if, as the lofty-worded preamble of the Japanese Constitution proclaimed, one could trust 'in the justice and faith of the peace-loving peoples of the world'. The outbreak of hostility on the Korean Peninsula in June 1950 put an abrupt end to that 'phoney' peace. In order to fill a gap created by the massive re-deployment to the Korean battlefield of US forces then stationed in Japan, MacArthur ordered the Japanese government to organise a 75 000-strong Police Reserve Force and an 8000-strong maritime force. These gradually evolved into the Self-Defense Forces (SDF). Also under the terms of the United States–Japan Security Treaty, signed in parallel with the Peace Treaty in September 1951, the United States was granted the use by its land, air and naval forces of facilities and areas in Japan 'for the purpose of contributing to the security of Japan and the maintenance of international peace and security in the Far East'. Thus a question arose as to whether the SDF and its role in connection with the Security Treaty were compatible with the war renunciation clause of the Constitution.

It is not this chapter's task to give a full account of the lengthy and often meticulously detailed debates about the constitutional problems relating to the SDF during the past four decades or so. Rather we are interested here in the way in which Japan's defence and security policies have been influenced by these debates. There are two major points at issue: (a) whether the very existence of the SDF is constitutional; and (b) even if it is constitutional, whether or not there is any constitutionally set limit to the exercise by Japan of the right of self-defence.

As stated earlier, the architects of Article 9 did not take the extreme view that Japan should be totally stripped of the right of self-defence. Based on the commonly held interpretation of the

Kellogg-Briand Pact, wars of self-defence are not prohibited. But this then begged the question as to the manner in which Japan was expected to exercise that right, given that the provision of Article 9 also stipulates 'land, sea and air forces, as well as other war potential, will never be maintained'. According to Ashida, who introduced the aforementioned amendment, what is prohibited by this provision is the maintenance of these forces for *aggressive purpose* only. In other words, according to this school of thought, Japan is not allowed to maintain these forces for the purpose of settling international disputes, but is allowed to do so for the purpose of self-defence.

Prime Minister Yoshida at first cautiously dissociated himself from this school of thought but as time went on he gradually changed his position to one nearly identical to that of Ashida, saying that 'war renunciation does not mean relinquishment of the right of self-defense' (Prime Minister Yoshida's speech at the Diet, 23 January 1950). Nevertheless, Yoshida's other well-publicised statement:

> It would be unconstitutional to have war-fighting capability (*senryoku*) even for the purpose of self defense. Without amendment of the Constitution, Japan cannot rearm itself, which is not the policy of my government. (Prime Minister Yoshida, Budget Committee of the Upper House, 10 March 1952)

indicates his reservations, probably based on his personal conviction, regarding the role of the military in postwar Japan. To Yoshida the SDF was simply that, and not the 'military'. A lasting testament to this philosophic approach is that even in the 1990s the word *guntai* (military) does not exist in Japan's official vocabulary.

In short, the real issue is not whether Japan has the right of self-defence (nobody seriously questions this point) but rather what should be the limits to the means of Japan's self-defence. Legal debates have been a substitute for political debates. Those in favour of the smallest possible armament (or non-armament) have been fond of taking issue on the constitutionality of the SDF, thus forcing the pragmatists to defend their positions on a battleground of legal controversy. The following excerpt from the latest issue of the Defense White Paper is a typical example of this:

Since the right [of self-defence] is not denied [by the Constitution] the government remains firm in the belief that the Constitution does not inhibit the possession of the minimum level of armed strength necessary to exercise the right of self-defense. On the basis of such understanding, the government has adopted defense-oriented policy as its basic policy of national defense and has maintained self-defense as an armed organization, and has taken steps to improve its capabilities and to ensure its efficient operation. These measures do not present any constitutional problems.[9]

For obvious reasons, what constitutes 'the minimum level of armed strength' that is permissible under the Constitution has been a focal point of the debates. Admitting that '[t]he specific limit of the necessary minimum level of armed strength for self-defense varies depending on the prevailing international situation, the standards of military technology and various other conditions', the Defense White Paper offers four criteria shaping the government's present defence policy. These are that:

1. offensive weapons (such as ICBMs, long-range strategic bombers and offensive aircrafts) exceed the limit;
2. three conditions exist where the use of a minimum level of armed strength is permissible (the existence of an imminent and illegitimate act of aggression against Japan, unavailability of other means than resort to arms, and even then the use of arms should be confined to the minimum necessary);
3. deployment of troops to foreign territorial land, sea and airspace goes beyond the minimum limit necessary for self-defence; and,
4. the exercise of the right of collective self-defence exceeds the minimum limit and therefore is constitutionally not permissible.[10]

As an incidental point, these criteria, except for the first, are more concerned with geographical/operational limits than quantitative/qualitative limits to armed strength itself. However, before proceeding to the question of geographical/operational limits, the topic of the next section, let us conclude this discussion by highlighting the following two points. Firstly, Japan's war renunciation clause in its postwar Constitution was (and is) in line with the development of international thinking over the decades (or centuries?) concerning the problem of war. In this context, the Kellogg-Briand Pact, the League Covenant and the UN Charter are all notable expressions of that thinking. Secondly, if the post-Second World War world had been if not an ideal then at least the semblance of an ideal world as envisioned by the founders of the League and the UN, postwar Japan would not have been put in a dilemma with regard to defence policy. In reality, however, as the Cold War developed, a gap between real and ideal worlds began to widen. Under these circumstances, it was inevitable for Japanese policy-makers to incline towards the pragmatic/moderate rather than the idealistic/extreme interpretation of the postwar Constitution's war renunciation clause. But given the people's rather strong preference for the idealistic interpretation, the political cost that the government has had to pay has not been insubstantial. What then will be the effect of the Cold War's end on Japanese thinking regarding its peace Constitution?

III   THE ENDING OF THE COLD WAR AND THE PEACE CONSTITUTION

As explained in the previous section, the orthodox interpretation of the Constitution's war renunciation clause is that while the country's right of self-defence is not denied, the means by which that right is to be exercised is strictly circumscribed. This may be termed a 'minimalist policy'. One of the methods to implement this policy is to place a limit on the SDF's arms/equipment and personnel. The defence budget's 1 per cent to GDP ceiling and the principle of minimum defence (*senshu boei*) are the most notable examples of this approach. (According to the latter, the SDF is allowed no offensive weapons, although the distinction between defensive and offensive weapons is not always clear.)

This minimalist policy has served its various purposes rather well over the past four decades or so. The government could therefore calm, if not silence, its opposition at home as well as constantly suspicious neighbours abroad. The policy was also useful in avoiding unnecessarily provoking the Soviet Union, the most formidable latent threat to Japan's security. Last but not least, it was useful in focusing the country's energy towards economic development.

This policy has however been a source of constant irritation in United States–Japan relations. While Japan is protected by the United States, Japan would not help the Americans if and when they were attacked. This incompatibility between the Alliance parties to the right of collective self-defence is an important ingredient of the minimalist approach.

The United States–Japan Security Treaty, by referring to the UN Charter, Article 51, takes cognisance of the inherent right of individual and collective self-defence possessed by both signatories. But despite this, the Japanese government is of the view that the exercise of the right of collective self-defence exceeds the minimum limit and is thus constitutionally not permissible. According to the Japanese government this means that Japan is not allowed to use its armed forces for the purpose of assisting its ally even when under armed attack by a third party as long as Japan itself is not actually under an attack from that third party. There being little probability that the homeland of the United States will be attacked, a more relevant question is what the SDF is allowed to do in support of American forces engaged in activities 'for the purpose of contributing to the security of Japan and the maintenance of international peace and security in the Far East' (Article 6 of the United States–Japan Security Treaty). The minimalist approach also applies to this type of situation in that it would be unconstitutional for the SDF to provide even logistic support for US forces thus engaged.[11]

In this respect, it should be noted that in practical terms there is no way of placing a geographical limit upon the activities of US forces as long as their activities are regarded as necessary for the purposes of the Security Treaty.[12] In fact the Japanese government has never attempted to place such a geographical limit upon the activities of the US forces stationed in Japan in the entire history of the Security Treaty. This may be termed a 'maximalist approach'. A clear distinction has existed between the minimalist attitude towards the activities of the SDF and the maximalist attitude towards those of the Japan-based US forces. This is the way in which an apparently one-sided alliance system has been effectively managed in the past.

The unchallengeable economic and military power of the United States has been a prerequisite for the successful management of the United States–Japanese alliance. As American power began to erode around the middle of the 1970s, there was an inevitable increase in US pressure on the Japanese to level-up the 'minimum' limit. This pressure was precipitated by the perceived threat from the Soviet Union whose military presence since the mid-1990s had a significant impact upon the security environment in Asia and the Pacific. The Japanese government responded to these pressures by, on the one hand, stepping-up measures of defence cooperation with the United States (e.g. the adoption of Guidelines for US–Japan Defense Cooperation in November 1978 and Japan's September 1982 pledge to extend its sealane protection role to a range of 1000 miles from the Japanese coast), and, on the other, by introducing new devices with which to keep the Armed Forces within bounds (e.g. the adoption of the Outline of National Defense Program in October 1976 and the GNP 1 per cent ceiling formula in November of the same year).

What is particularly relevant to the present topic is the SDF's geographical scope necessary for its role in sealane protection. Although close collaboration with US forces is indispensable for such activities, the Japanese government takes the view that this is not a case of the exercise of the right of collective self-defence, basing its judgement on Article 5 and not Article 6 of the Security Treaty.[13]

In short, the United States–Japan Security Treaty would not be useful for 'out of area' security cooperation as long as the Japanese government adheres to a minimalist interpretation of the right of collective self-defence. The weakness of this arrangement had not been exposed during the Cold War era because: (a) the Americans did not really expect a significant security role for Japan, being content with the privileges the United States was granted for basing rights in Japan; and (b) Japan, being concerned over probably disastrous complications with the Soviet Union, had been effectively deterred from playing any significant security role in such remote areas as the Middle East or even in Indochina. The minimalists are still firmly holding their ground, referring, among other things, to the negative response from neighbours to a proactive stance by Japan. With the disappearance of the Soviet threat, however, the American public will grow more and more impatient of the minimalist view of the Security Treaty. Such a view is also likely to be regarded by world opinion not as a sign of Japan's commitment to international peace but rather as a sign of a self-centred attitude.

One possible solution to this dilemma is to reinterpret the peace Constitution. One of the basic premises on which the war renunciation principle was founded was the undisturbed operation of a UN collective security system. Unfortunately, the Cold War meant that the UN's security role was suppressed. As a result, there developed a widening gap between the ideal of Article 9 of Japan's Constitution and the reality of international life. The ending of the Cold War will not immediately solve this problem but it will certainly help Japan to narrow the gap between the ideal and reality. Is it not high time for Japan to make a more 'fruitful' use of its peace Constitution for the furtherance of peace enforcement, peacekeeping and peacemaking through the United Nations? Japan's active role in such endeavours is undoubtedly in line with the constitutional principle of war renunciation because the UN Charter and the postwar Japanese Constitution are rooted onto the same principle. Japanese thinking about international security should be encouraged to develop along this line and should not be distorted by the minimalist interpretation concerning the right of collective self-defence. These two concepts – collective self-defence and collective security – should not be confused.[14]

Notes and References

1. See, for example, Yamauchi Toshihiro, 'Gunning for Japan's Peace Constitution', *Japan Quarterly*, April-June 1992, pp. 159–67. He is the author of *Heiwa Kempo no Riron* (Theory of the Peace Constitution) (Tokyo: Nihonhyoron-sha, 1992). See also, Higuchi Yoichi, 'When Society is Itself the Tyrant', *Japan Quarterly*, October-December 1988, pp. 350–6, in which the author discusses a case in which the wife of an SDF (Self Defense Force) serviceman, who was killed while on duty, sued the Veterans' Association that had requested enshrinement of the dead at Yasukuni Jinja (i.e. a Shinto shrine). The issue was primarily concerned with separation of church and state, but implicitly related to the constitutionality of the SDF itself.

2. Osamu Nishi, *Ten Days Inside General Headquarters (GHQ): How the original draft of the Japanese Constitution was written in 1946* (Tokyo: Seibundo, 1989).

3. Osamu Nishi, *The Constitution and the National Defense Law System in Japan* (Tokyo: Seibundo, 1987) p. 9. Nishi quotes Kades from his interview with him on 13 November 1984. See also, Nishi, *Ten Days*, pp. 36–41 for Kades's role in the war-renunciation clause.

4. Nishi, *The Constitution*, ibid.
5. Hitoshi Ashida, *Ashida Hitoshi Nikki* (Diary), vol. 7 (Tokyo: Iwanami shoten, 1987) pp. 298–301 and 318–20.
6. Nishi, The Constitution, p. 10.
7. Ashida wrote in his diary that, Japan being a signatory to the Kellogg-Briand Pact and the League Covenant, the idea of war-renunciation and pacifist settlement of international disputes itself was nothing novel to the Japanese, vol. 1, p. 80.
8. It is interesting to note that the interpellator in question was Sanzo Nosaka, leader of the Communist Party of Japan. His argument ran as follows: We must tenaciously maintain the independence of our nation. The Communist Party of Japan is determined to sacrifice everything in the fight for our nation's independence and prosperity. There is the risk that Chapter Two (Article 9) of the Constitution will renounce our country's right of self-defence and threaten the independence of our nation. For this reason our party must oppose this Constitution for the sake of the independence of the nation. Nishi, *Ten Days*, p. 195.
9. Japanese Defense Agency, *Defense of Japan 1991*, pp. 54–5.
10. Ibid, pp. 55–6. This document maintains, in addition to the four criteria mentioned in the text, that the use of the minimum force necessary for self-defence is different from exercising the right of belligerency, which is prohibited by the Constitution. This is another moot point inherent in Article 9. Until recently, a further device had been in use to determine the minimum level of arms, namely the ceiling of 1 per cent of GNP for the annual expenditures on defence.
11. Shuzo Hayashi, Chief of the Cabinet Legislation Bureau, Budget Committee of the Upper House, March 1959.
12. The government's coordinated view on the treaty area (Far East) as it is stipulated in the US–Japan Security Treaty, presented before the Special Committee on Security Treaty, Lower House, 26 February 1960.
13. Director of Self-Defense Agency, Michita Sakata, at the Budget Committee of the Upper House, 12 June 1975. Article 5 of the treaty reads as follows: 'Each Party recognises that an armed attack against either Party in the territories under the administration of Japan would be dangerous to its own peace and safety and declares that it would act to meet the common danger in accordance with its constitutional provisions and processes.' Any action that the SDF may take under the provision of Article 5 is regarded as the exercise of the right of *individual* self-defence even if it actually acts in conjunction with US forces. Is it possible, however, to distinguish the activities of the US forces that will be taken under this Article from those under Article 6?
14. A report submitted by an LDP task force headed by Ichiro Ozawa, formerly Secretary-General of the Liberal Democratic Party, takes a similar position on this issue. See the Ozawa Report, *Kokusai shakai ni okeru nihon no yakuwari* (Japan's Role in the International Society), *Bungeishunju*, April 1992, pp. 132–45. By emphasising the importance of Japan's positive role in the collective security functions (broadly defined) of the UN, the author does not intend to ignore the importance of another problem – the need to adjust the minimalist interpretation of the right of collective self-defence to

the changing reality of US–Japan relations. In his view, a more flexible approach to the collective self-defence issue (which is related to the *modus operandi* of US–Japan Security) and a more proactive approach to the UN collective security system are not mutually exclusive. The emphasis is that *both* are necessary. See also, Masashi Nishihara, 'New Roles for the Japan–US Security Treaty', *Japan Review of International Affairs*, vol. 4, no. 1, Spring/Summer 1991, pp. 25–40.

3 Civilian Control of the Military in Postwar Japan
Ian Gow

One of the first scholars to attempt a rigorous theoretical framework for civilian control of the military was Samuel P. Huntingdon.[1] He created two opposing concepts, subjective control and objective control, and these have been refined and also at times criticised by later scholars.[2]

By objective control Huntingdon meant the achievement of control by militarising the military, making it a tool of the state. He therefore held that objective control conferred a maximisation of military autonomy which in turn would create a professionalism which would preclude or limit military intervention and by definition enhance civilian control, or at least obviate the need for civilian control. By subjective control Huntingdon meant the maximisation of the power of one or more political groups (such as a governmental institution or party) over the military, making the military the mirror of the state. He clearly believed that subjective control would deny the existence of an independent sphere of purely military objectives.

What seems to have gone unnoticed is that Huntingdon's civilian control concepts seem to accord well with the actual frameworks existing in prewar and postwar Japanese civil–military relations. Thus the case of modern Japan's civil–military structure may indicate the potential dangers for effective civilian control and the prevention of military intervention or role expansion in politics inherent in either of Huntington's ideal types. Japan therefore, especially when one contrasts the prewar and the postwar system, offers a fascinating set of case-studies for scholars of civilian control of the military (or civil supremacy over the military). For example, the objective control created deliberately in prewar Japan led inexorably to massive military role expansion in politics under the aegis of national security and, especially after the First World War, total war planning. The postwar *Bundeswehr* (and the Japanese *Jieitai* to some extent) have, it has been claimed, suffered from excessive civ-

50

ilianisation or permeation of civilian values to the detriment of military effectiveness. This has been commented on in early postwar debates on the lack of *'inner fuhrung'* or *yamato damashii* etc., and thus cast some doubt on how far down the system the civilian control processes should be permitted to penetrate or permeate. Thus the 'conservative' objective control and 'liberal' subjective control still offer analytical insights which can, irrespective of perceived flaws in the original concepts, assist us in understanding the modern Japanese civil–military structure.

In order to better comprehend the problems facing today's Japanese policy-makers regarding civilian control, this chapter will provide first an overview of civilian control, military insulation from political control and military role expansion in domestic and foreign policy-making processes in the prewar period. This is necessary in order to be able to place postwar issues and fears of an increased military role and expanded autonomy in context. Next the origins of the postwar system, their evolution and the formal frameworks will be described. Finally, the recent heated domestic political debate over Japanese participation in United Nations 'Peacekeeping Operations' (hereafter PKO) will be used as a focal device for indicating both the present state of civilian control and the problems of changing the existing system. Since this issue culminates in a major political debate over the role of the Japanese Parliament (Diet) in controlling the dispatch of Japan's Self Defense Forces overseas, it is particularly apposite as an example of the continuing problems and sensitivities surrounding the present Japanese concept of civilian control.

I THE PREWAR SYSTEM

Prior to the establishment of the modern state (1868) the Japanese, for at least two hundred years, had resolved their civil–military relations by creating a unified, centralised system, the *bakufu* (lit. Camp Government) under a military leader called the Shogun (lit. barbarian-repelling generalissimo). The Shogunate, during the 200 years of national isolation (*sakoku seisaku*), went through a gradual metamorphosis from the military arm of the ruler (the Emperor) into a civil–military oligarch who actually ruled the country whilst the Imperial Institution was reduced to a mere cipher.

The fall of the Shogunate and the emergence of the modern form of governmental structures with the Imperial Institution, placed again at the very centre, occurred in the late 1860s due to a combination of internal and external pressures on the system. The new system provided for a military establishment which was actually subordinate to the Prime Minister equivalent (*Dajo Daijin*) of the day.[3] Influenced by the French system, the military leaders of the Hyobusho (Ministry of Military Affairs) reported to the Emperor through the Premier and not directly as had the Shogun in the past.

In 1878, at the instigation of Yamagata Aritomo, hero of the Restoration War (1867) and father of the Imperial Army, a general staff system was created based on the German system whereby the Chief had direct access to the Emperor and this was permitted in peacetime as well as wartime (the German model was for wartime only).[4] This single organisational change drastically altered the nature of civil–military relations in Japan for the entire prewar period. It provided the Chief of the (Army) General Staff on matters of command (*gunrei*) with the ability to bypass not only the Premier and his advisers but also the uniformed Army Minister (Minister of War) and petition the Emperor directly. This right of direct access (*iaku joosoo*) was eventually extended (after 1890) on administrative matters (*gunsei*) to the Army and Navy Ministers, as well as the Chief of the Naval General Staff and often other commanders.[5] In addition, in wartime the Army General Staff controlled the Navy through Imperial Headquarters. Yamagata's main reason for doing this was apparently to insulate the military from 'interference' from 'politicians' especially the emerging party politicians. In effect, however, it laid the foundations of 'dual government' whereby military matters were decided by military authorities alone. At this time too the practice of appointing only serving officers as army and navy Ministers began to be practised. Thus, prior to the establishment of the cabinet (1885), the Constitution (1889) and the Parliament or Diet (1890) the military had already begun to insulate itself from political control. Any attempt to intervene excessively in military matters (including budgets) would result in the refusal of the military authorities to appoint a service minister to the Cabinet or even withdraw a minister and thus bring the Cabinet down. This insulation of the military from political control or supremacy was further strengthened by allowing the Diet only very limited powers over the military budget (if it vetoed the present year's budget then the previous year's allocation was automatically approved). In 1900 a new set of regulations was established (also called *gunrei* or military commands) which permitted strategic transfers of powers between the ministries and general staffs without cabinet approval. Thus, as happened in the 1920s, when there were attempts to civilianise the service ministry portfolios, the powers were capable of being transferred to the general staffs.

Thus, by the time of the Russo-Japanese war, the military were effectively insulated from civilian control formally. The real power in this period, however, did not lie with Parliament or the cabinet. Rather the political system and especially civil–military relations were actually controlled by an extra-constitutional body called the *Genro*, comprising leading high ranking figures in the Imperial Restoration who decided on the Prime Minister etc., and actually ran the country. This elite group of elders, since it comprised men with extensive military experience, was actually a civil–military oligarchy. This enabled the *Genro*, especially the 'military *genro*' Field Marshall, Prince Yamagata Aritomo, to manage the military effectively.[6] However, by the end of the first decade of the twentieth century the *Genro* were openly divided, weakened by internal rivalries and decimated by ill health and death from old age etc. Therefore the control system began to deteriorate rapidly just as the major problems in civil–military relations were beginning to emerge.

II MILITARY ROLE EXPANSION

The Imperial Rescript to Sailors and Soldiers (1882) specifically forbade active service personnel from involving themselves in politics. However, since service ministers had to be, in practice, active serving officers, these key cabinet positions provided a continuing opportunity for service personnel to influence policy if not actively intervene. This was especially true on national security matters. As Japan moved from 'cordons of sovereignty' to 'cordons of interest', national security began to encompass major aspects of foreign policy and even domestic policy.[7] After the First World War the requirements of Total War planning were to provide, under the aegis of national security, extensive opportunities for the military to declare that all aspects of economy and society were open to military involvement and thus military role expansion in politics increased. In addition, the continuing budgetary battles for expanding army and navy needs were exacerbated from the Russo-Japanese war onwards by the designation of the Russians as the army's potential enemy and the Americans as the navy's target. Since the *Genro* was less effective, and since there was no real effective coordinating mechanism other than the Imperial Institution (which had to remain 'above politics') the two services began to attempt to influence the political system outside cabinet, etc., thus increasing military role expansion in politics even more.

There were only two really major efforts to control the military. The first arose immediately prior to the First World War to resolve these paralysing budgetary battles over national defence plans which threatened to totally undermine the cabinet system. The second occurred in the interwar period from 1921 to 1931 through developments related to the international naval arms limitation initiatives.[8]

In 1912 Premier Admiral Yamamoto Gombei, in an effort to break the cycle of collapsing cabinets over competing army and navy budgetary demands (the so-called Taisho Political Crisis), proposed and actually obtained approval for new regulations which would permit reserve officers to serve as service ministers. This would enable cabinet positions to be allocated to officers, no longer subject to military discipline. This was, in fact, a hollow victory in that no reserve officer ever occupied the service minister portfolios[9] and in the 1930s the regulation was discarded. These efforts at civilian control by an Admiral, however, question the conventional wisdom amongst Western scholars following the interpretation contained in the work of Maxon, that military officers as Prime Ministers were an indication of military control. In many cases they were an effort by the civilian elite to try and control the military by utilising former military leaders.[10]

During and immediately after the Washington Conference (1921–2) the navy and the civil authorities flirted with suggestions of a civilian service minister structure, only to trigger off developments to strengthen the Naval General Staff.[11] Some progress was made in controlling naval and army expenditures but it actually overstrained the fabric of civil–military relations and probably contributed to the politics of violence and intervention by young officers and military elites respectively in the 1930s. During the Washington Conference (1921–2) and London Conference (1930) the Prime Minister (on both occasions a civilian) actually occupied the navy ministry position pro tem. This was regarded as a major slight on the military prerogative by the extreme right wing and both Prime Ministers were in fact assassinated (by civilian fanatics).

Budgetary politics enhanced the possibility of political intervention and the prevention of cabinets being formed or the pulling down of cabinets for failure to meet the competing and ever increasing budgetary demands of the army and navy. The 1910 National Defense Plan, with the potential enemy for the army being Russia and for the navy being the United States, was a blatant overt example of the likelihood that competing service budgetary needs would necessitate military intervention in a civilian budget-making process. After the First World War the problem was exacerbated by military expansion on the Asian Continent. The expansion of the Army in Siberia after the First World War, in China in the 1920s and in Manchuria in the 1930s, showed how incapable civilian elites were of controlling military expansionism at home or abroad and were a major factor pulling Japan into a direct conflict with the Anglo-American powers. Whilst it would be going too far to state that the Japanese military totally controlled Japanese society in the 1930s or even during the Pacific War, it would be true to say that they were very effectively insulated from civilian control and political intervention by civil authorities whilst at the same time more than capable of actively intervening in 'civilian matters' under the aegis of national security. The major problem was, that since the military officials could directly access the Emperor, bypass or block the cabinet etc., the only coordinating mechanism was the Emperor himself, the Commander-in-Chief. But he was above politics and was roundly threatened whenever he or the Imperial Institution did intervene (which was rare). The Palace, through the officials of the Imperial Household Agency did, in fact, act as a device for civilian control especially during the London Naval Conference and the so-called Crisis of the Infringement of the Supreme Command (*Tosui ken Kanpan*

Mondai) in 1930. However the victory for the constitutional monarchists over the right wing and military party led to them being subjected to violence and a major shift, even within the Palace itself, in the 1930s.

III POSTWAR DEVELOPMENTS

In 1945 Japan was placed under Allied, essentially American Military Government Supreme Command Allied Forces (SCAP). The Allies were determined to disarm, demilitarise and democratise Japan. One immediate target was the role of the military in postwar Japan. The key document emanating from Washington was SWNCC 228 but this simply stated that the 'civil be superior over the military', hardly surprising given the nature of dual government and military autonomy in the prewar period. Japanese policy-makers were instructed to draft new clauses and these were essentially similar in that they envisaged a postwar military but one strictly controlled by, and subordinate to, the civilian government. SCAP planners in Japan itself, however, extremely dissatisfied with the Japanese revisions, then drafted a new constitutional clause which implied that there would never again be a military and this was subsequently revised a number of times by both SCAP officials and the Japanese government (Ashida Amendment). The authorship of the clause has been a matter of some debate, but given the similarities in wording to the prewar Kellogg-Briand Pact, it seems more likely that Shidehara (Foreign Minister at that time) rather than General MacArthur, was the author. The resultant clause shifted from the preface to Article 1 and then became Article 9; the famous 'peace Clause'. The wording became less precise and the Allied Council in Japan, fearful of the loopholes now evident in the clause and aware that it could now be interpreted as permitting a future military, insisted that Article 66 be amended. Reiterating the point made by Professor Watanabe in the previous chapter, the new Article 66 included a clause stating that 'all cabinet/State Ministers must be civilians'. This, in fact, required the Japanese to coin a new word (*bunmin*) for the term 'civilian' since hitherto this had not existed in the Japanese language. Over the next few years this 'civilian' clause was often (mis)interpreted to preclude former career officers

from occupying high government posts but in fact it was a key
constitutional innovation which provided for an effective civilian
control system and precluded cabinets being dominated by military officers as in the prewar period. During the Occupation
(1945–52) there was, in addition, a major purge of career officers
who were effectively precluded from holding any public office, a
temporary demilitarisation and civilian control measure. In the
late 1940s the major debate was not about disarmament but
whether Japan could effectively be protected by the United
Nations or by the United States. The intensification of the Cold
War eventually dictated that it be the United States. For the first
four years of the Occupation there was no public debate about
rearming Japan but there was emerging an American consensus
that Japan should provide support for US aims in the Far East
and even a small rearmament to share the burden with the
United States within Japan and in the Asia-Pacific Region. The
Japanese Imperial Forces had, of course, been disbanded,
although units of the Japanese navy continued under UN/
American control even during the Korean war, on minesweeping
duties. However the need to dispatch US forces based in Japan
to Korea caused MacArthur to order the establishment of a
substantial National Police Reserve (NPR) (*Keisatsu Yobitai*).[12]
Portrayed, even to those joining it, as a police force, it was in fact
a nascent army. At first ex-military personnel were disbarred
from the new force but this was quickly discarded and former
officer trainees of the now defunct Imperial officer training
schools were allowed to join. The creation of a National Safety
Force (*hoantai*) and then finally in 1954 a Defense Agency
(*Boeicho*) and tri-service Self Defense Force (SDF) (*Jieitai*) replaced the NPR. This force, under American tutelage and expanded gradually under American pressure, had initially a very
limited role in external defence and a support role within Japan
itself, with the main burden falling on the US forces based in
Japan under the Mutual Security Treaty (1951). However, con-
tinual American pressure has resulted in Japan agreeing to
sharing more of the costs of US forces (on a much reduced level)
based in Japan and a gradual shift to a larger role in Japan's
external defence and the main burden for defence of the homeland resting now with the SDF.

Japan's forces, developed within a policy framework, severely
constrained by interpretations of Article 9, are today the most

technologically sophisticated non-nuclear force in the Asia-Pacific. It could be argued, however, that it is a different kind of armed, uniformed force, beyond para-police or para-military but essentially defensive. Indeed for nearly three decades Japanese politicians in power avoided ever using the term *guntai* (military) to describe the SDF.

IV CIVILIAN CONTROL (*BUNMIN TOSEI* OR *SHIBIRIAN KONTOROORU*)

Leaving aside extreme interpretations of Article 9 which would effectively preclude any military at all (the ultimate civilian control), what are the formal mechanisms which ensure effective political superiority over the military? The Commander-in-Chief (the PM, not the Emperor as prewar) must be a civilian (Article 66). The Director General of the Defense Agency, as a State Minister, must also be a civilian by the same Article. The Defense Agency has never been upgraded to a Ministry and the Director General is therefore not a member of cabinet as of right. Thus the Defence Ministry was kept weak and due to its sensitivity it also did not attract high-calibre politicians, rendering it easier for other elite ministries such as Foreign Affairs and Finance to control military matters. There was also, from 1955 to 1985, an all-civilian National Defense Council (*Kokubo Kaigi*) chaired by the PM, although in actuality it seems to have been a rubber stamp for Defense Agency decisions. Efforts to develop comprehensive security policy in the 1980s resulted in the creation of a Ministerial Council on Comprehensive Security within the Cabinet, again all-civilian. This was an effort to achieve better coordination and planning through a more effective mechanism than the National Defense Council after the government had failed to reorganise, reposition and strengthen the latter.[13] In 1989 the National Defense Council was finally reorganised into a National Security Council and this was again all-civilian. The highest uniformed member of the SDF, Chief of the Joint Staff Council (*Bakuryo Kaigi-cho*), is not a member but can be summoned to give evidence/advice. Japan's Parliament (Diet) has never really developed an effective system of civilian control since it basically could not decide, at least formally, through debates covering almost 25 years, whether the SDF was a military or not. From time to time suggestions of uniformed officers giving evidence in the Diet caused a furore over whether this was permissible or not. In 1980 *ad hoc* Committees in both Houses on defence-related matters were, for the first time in the postwar period, established, but these were consultative and advisory rather than policy committees.[14] The only area where the Diet arguably formally exercises any real control is in the dispatch of SDF forces. The PM is required to have permission, or, if the Lower House is not in session, obtain it as soon as possible thereafter. The only area where the Opposition parties in the past have managed to attempt some form of control is in the budgetary debates but this has always been relatively ineffective.

There are a number of interpretations of various clauses of the Constitution, including, but not limited to, Article 9, which enhance civilian control and arguably make the SDF a rather different type of armed force. There is no system of military justice since Article 76 of the Constitution is at present interpreted to preclude court martials. Conscription is interpreted as unconstitutional under Article 18. There is no national security law and attempts to create one have met with strong resistance by the public and the opposition. Front-line commanders require permission before firing on an enemy, even if that enemy has landed in Japan. One top officer, Kurisu, was dismissed for suggesting that he would not wait for authorisation if the enemy landed. Article 9 is interpreted as implying that only defensive defence is possible, and dispatch overseas too is precluded (but see below). Cabinet resolutions relating to the 1 per cent ceiling on defence (recently exceeded), on the three non-nuclear principles and on the export of weapons, are all forms of civilian control but are reversible without constitutional amendment since they are Cabinet decisions based on the spirit of Article 9. Two terms are normally used in the literature to describe this formal array of restrictions on the SDF, *Shibirian Kontoororu* and *Bunmin Tosei*. However, less well known but arguably more effective (and possibly detrimental from a military command perspective) is the formidable civilian control system operated within the Defense Agency known as *Bunkan Tosei* (bureaucratic control).

V   BUREAUCRATIC CONTROL (*BUNKAN TOSEI*)

In the National Police Reserve (NPR) era (1950–2), only civilian former police or Home Ministry officials advised the uniformed Commandant directly; uniformed staff were excluded and were in effect advisers to the NPR bureaucrats. However, the National Safety Force (1952–4) actually offered equal access to the top from both the uniformed officers (similar to *gunrei* – military command issues prewar) whilst administration (*gunsei*) was the preserve of the civilian officials. In 1954 with the creation of the SDF the internal bureaux were greatly strengthened and the leading bureau chiefs either had a Home Office or police background or were seconded from elite ministries. These internal bureau (*naikyoku*) chiefs controlled key areas such as strategic planning, promotions above Captain, etc., and were extremely powerful. Moreover only they were entitled to offer advice direct to the Director General, and thus, combining both administrative and political power in this way were, and are, civil servants with extensive powers beyond the norm in Japan. The Chairman of the Joint Staff Council, the highest uniformed officer, in theory advises the Director General of the Defense Agency (not the C-in-C, the Premier) but in practice is merely an adviser to the internal bureaux chiefs. His best access to the Director General and/or the Commander-in-Chief (PM) is indirectly through the United States–Japan Security Treaty Committees via US officials, uniformed and civilian. Whatever the euphemism used within Japan, previous Chairmen of the Joint Staff Council, Kurisu and then Takeda, were dismissed for trying to give the uniformed force a far greater public role and to influence policy more directly by somehow bypassing the extensive bureaucratic control system by the internal bureau chiefs.

The system of bureaucratic control, whereby civilian officials trained often to regard SDF matters as police or Home Ministry related, is probably excessive. Secondment of elite bureaucrats to key internal bureaux may, of course, preclude Defense Agency bureaucrats becoming too influential, since the latter are liable to 'agency capture' by their uniformed colleagues. However it does create a 'glass ceiling' in terms of real promotion for Defense Agency bureaucrats. More important, however, military intelligence and strategic thinking is being filtered through these civilian bureaux chiefs and this may actually impair military effectiveness or effective policy-making by the elected civil officials responsible. In the end, of course, one would not argue against civil officials and politicians disregarding purely military advice but the Director General and the C-in-C should have both sets of advice and then decide.

Civilian control mechanisms in Japan do exist and in many respects appear very effective. The fact that Nakasone achieved the Premiership after being Defense Agency Director General meant that this post was no longer so sensitive (or powerless) that able and ambitious politicians would avoid it. This in turn can be seen as enhancing civilian control although a weak politician in post may allow this agency to be dominated by other elite ministries, a form arguably of bureaucratic and therefore civilian control. The Diet should have Standing Committees not *ad hoc* Committees dealing with security matters, and might even, on the American system, approve the highest appointment of the Joint Staff Council.[15]

VI THE GULF CRISIS AND CIVILIAN CONTROL

The problems Japan faced in dispatching forces to the Gulf in the war against Iraq were, of course, a key issue in United States–Japan relations and in Japan–UN relations. They also throw considerable light on the problems of coalition-building within Parliament over security-related legislation. However, at base are the key issues related to civilian control, such as overseas dispatch, and these will be the focus here.

Dispatching SDF forces overseas, no matter what the reason, has always been problematical in postwar Japan. In large part it is due to the Japanese memory of an uncontrolled military expansion in Siberia and Manchuria prewar, and also Southeast Asian nations' fears of a postwar revival of Japanese militarism emanating from a Japanese military presence outside Japan (even for disaster relief), again a memory of the prewar and wartime period. It is often erroneously stated that there is a constitutional ban on overseas dispatch. This is not true, although one could argue that it contravened the spirit, if not the wording, of Article 9. Prior to the June 1992 Diet vote to allow Japan's military to participate in UN Peacekeeping operations, it was illegal to dispatch armed forces abroad. But the legal impediments were interpreted as being over the SDF Laws rather than the Constitution. Since the SDF laws did not state that the SDF could be dispatched overseas, then any action to carry this out would have required at minimum amended regulations, otherwise the act was illegal. However, Japanese SDF staff prior to 1992 had been sent abroad for training and on diplomatic duties as defence attaches. This is normally referred to as '*hakken*' (dispatch) and is contrasted with '*hahei*' (armed dispatch). The inclusion of Japanese forces in the training exercises for RIMPAC (Rim of the Pacific – a naval training arrangement between Australian, New Zealand and US forces), although sometimes seen as an infringement of the ban on overseas dispatch, is actually more problematical in terms of potential infringement of the Article 9 prohibition of involvement in collective security.

Given the legal constraints and the political sensitivity of overseas dispatch, both within Japan and amongst Japan's neighbours, the Gulf crisis, and indeed previous oil crises, have repeatedly presented Japanese policy-makers with a major dilemma. This was further exacerbated, in terms of the Gulf War, by the fact that the initial force deployment was not a sanctioned UN force but an American-led anti-Iraq force hoping for UN sanction. The idea of Japan participating in such a force prior to UN-sanctioned approval caused yet a further problem for Japanese policy-makers, since they feared that this would provide the means for uniting the opposition over the dangers inherent in the United States–Japan Security Treaty. For almost the entire postwar period the great fear of the Japanese socialists and others on the left had been that the United States–Japan Security Treaty would drag Japan into direct armed conflict. This had not occurred during the Korean and Vietnam wars and had undermined the leftist coalition's position on this matter. In addition, Article 9 does clearly ban the use of force to settle international disputes. Once the United Nations had sanctioned the US-led armed coalition the Japanese debate moved to a different level and focused on participation in the fighting (peacemaking) versus participation in the peace settlement (peacekeeping). Article 9 is certainly normally interpreted to forbid involvement in collective security but some Japanese scholars would argue in two ways against this.[16] The first is that UN operations are not classified as collective security. But, more important, they could argue that not participating in the UN action was itself unconstitutional. Article 98 of the Japanese Constitution states that 'Japan shall honour all treaties'. Japan has signed and ratified the UN Treaty Agreement and, within that, Article 43 states that 'all members make available to the Security Council, on its call, armed force – assistance and support'. In addition UN Article 4 states that:

> In order to contribute to maintain international peace and security all members shall promise to let the Security Council use necessary troops, aid and convenience for the maintenance of international peace.

However, assuming as some scholars do that international law transcends domestic law, the UN regulations are held to apply to a UN Force not a US-led armed coalition sanctioned by the UN.

In August 1990 a government proposal for a UN Peace Cooperation Law was put forward which would exclude SDF forces altogether. However a shortage of volunteers and the problems of training and readiness meant that the Government then attempted to include SDF members. These were initially to comprise medical and other officers from the SDF, and the debate focused on whether men were sent as individuals or as units and whether they could carry arms for other than basic self-defence. This proposal, though initially supported by two of the major opposition parties, the Komeito and the Democratic Social Party (DSP), then lost their support. They perceived the original agreement as being premised on zero participation by the SDF but had witnessed a gradual shift via a marginal support role for the SDF to a central (and sole) role for the SDF. They were now to be permitted sidearms and would wear SDF uniforms with the addition of an armband. The issue split the ruling Liberal Democratic Party, caused the defection of the two opposition parties, triggered off tremendous opposition from the anti-war movement in Japan and alienated Japan's Asian neighbours. Despite the fact that the bill was to include a clear statement that this force would never use force or the threat of the use of force, the bill failed to make it to Parliament but the opposition parties agreed to reconsider the bill on the proviso that it be significantly altered.

In the next phase, in late 1990, the LDP again managed to

achieve a working consensus with the Komeito and the DSP and even the normally hostile Japan Socialist Party agreed to 'discussions only': a major breakthrough. However, the new bill was again to exclude the SDF. By this time of course the major conflict was over, and in April, following the decision by West Germany (which was facing similar problems on overseas dispatch) to send minesweepers to assist in the clearing-up of mined areas, Japan also concurred and dispatched minesweepers. The government, having suffered major embarrassment internationally since it could only offer money rather than 'sweat and blood', continued to negotiate the new UN Peace Cooperation Bill. By August 1991 the bill not only did not preclude the SDF, it actually centred on sole SDF participation. The SDF were to make up the actual special force (2000) to be dispatched. These forces were to be dispatched, however, only under five clearly agreed conditions:

1. That there be a ceasefire in place.
2. That the Japanese force has the agreement of the parties involved.
3. That strict neutrality be maintained.
4. That the Japanese force would withdraw if conditions were not met.
5. That the use of arms be for self-protection only.[17]

This special corps within the SDF, including the establishment of an HQ within the PM's office, actually looked as if it might, due to Japan's acute international embarrassment, gain sufficient support amongst the opposition parties to enable passage of the UN-related bill through Parliament. However, what was emerging was a kind of consensus that the SDF could participate in UN peacekeeping duties but not a UN Force engaged in peacemaking. Presumably Japanese policy-makers were looking to participation in UN peacekeeping as an incremental gain towards full overseas dispatch in the future. The government also indicated that it was considering SDF participation in international disaster relief (another incremental gain for overseas dispatch) at the timely moment the minesweepers arrived home having been dispatched on peacekeeping duties without any untoward events and without having created any furore at the time of their dispatch. However, the Japanese government, poised on the verge of a major breakthrough, incomprehensibly shattered the consensus over, of all things, prior parliamentary approval of dispatch and retrospective sanction of the dispatch of the forces in an emergency. The DSP in particular, a key player in the Diet coalition required to ensure passage of the UN law through the Lower House, insisted on prior approval. This was consistent with the Premier's actual powers *vis-à-vis* the SDF in general regarding domestic dispatch. The DSP demand was initially brusquely rejected, and then the government, offering concessions, severely qualified these. The Cabinet seemingly were only prepared to submit for sanction after the event and it wished the debate to focus on the period (2 years down to 6 months). Ironically, then, the Diet was challenging the Prime Minister's right to dispatch without Parliamentary approval, the only area in which the Diet historically had any real control over SDF domestically. Admittedly the special force was within the Prime Minister's office and also there was a precedent for the PM to act without Diet approval in the period 1950–4. In December of 1991 the LDP-dominated Research Commission on the Constitution rejected a request to amend Article 9 to permit the SDF to participate in UN military missions, and it seems that once again the issue would be shelved, but the LDP made one last major effort throughout early 1992 which culminated in major breakthroughs permitting dispatch on peacekeeping (but emphatically not peacemaking) UN operations and dispatch on international disaster rescue operations.

The Gulf crisis, in effect, whilst almost ending in defeat for the government, actually provided the necessary impetus to make a real breakthrough in terms of overseas dispatch. After almost 20 months of intra-party negotiations the LDP finally managed to create a tripartite coalition with the Komeito and the Democratic Socialist Party and pushed two bills through the Upper and Lower Houses. The bills were rigorously (and in a highly emotive way) opposed by the Socialist Democratic Party and the Japan Communist Party, but although the pasage of the bills was delayed and some alterations made in the Upper House to strengthen civilian control, in June of 1992, the so-called PKO Law – really the International Peace Cooperation Law was approved. Dispatch of forces on UN duties, on condition that the missions were endorsed by the UN, agreed by the warring parties involved, and that immediate withdrawal would occur if the ceasefire was broken, was now legal. Japanese SDF personnel now began to ready for non-combat duties including logistical support, medical assistance and election monitoring and these were termed non-PKF duties in the new legislation. Units of the SDF have now been sent to Cambodia. Undoubtedly the government would have preferred a first dispatch outside of Asia where memories of the Second World War still linger but the domestic and foreign backlash which was feared did not really materialise. However SDF participation in more hazardous operations, which might require SDF personnel to use small weapons in self defence, will require Diet approval and additional legislation. These are specified in the legislation as PKF duties. If new legislation is put through for PKF activities it has already been agreed that the Premier would require Diet approval and the Diet must decide within seven days. In addition if the SDF is dispatched on PKF duties the premier must ask for Diet approval to extend the dispatch. Finally, the new legislation is to be reviewed by the Diet every two years. This has undoubtedly eliminated the main objections of the two minor parties, DSP and Komeito, which insisted on stronger civilian (parliamentary) control over the dispatch. The one loophole in the existing legislation perhaps is that the restrictions on despatch refer to troops, leading some to speculate that the SDF may now build up expertise in all UN-related areas of PKF by dispatching officers as opposed to men.

This legislation has undoubtedly eased Japan's growing international embarrassment, particularly since UN-related peacekeeping activities around the world look set to increase. For the opposition and a declining section of the public and even some of the nations of Asia, the new legislation may present a threat rather than an opportunity. The presence of Japanese troops on foreign especially Asian soil, for these people spells the return of Japanese militarism and the beginning of the end of civilian control. For others, however, and they now seem the majority within and without Japan, the new legislation, including the Foreign Ministry-inspired International Emergency Assistance Law permitting SDF despatch on emergency disaster relief operations, is not a threat at all. Rather, it indicates that civilian control and the sovereignty of the Diet is now reasonably assured. In addition, it represents a painful but highly medicinal dose of political realism to the body politic in the areas of defence policy and international relations. In the end the LDP gambled on the age-old Japanese strategy of using external pressure (*Gaiatsu*) to advance a particularly problematical domestic problem. The fact that it was UN-related rather than directly US-related probably made it that much easier and the LDP does not appear to have suffered any negative effects in recent electoral competitions. Whether this new breakthrough represents the thin end of the wedge for a movement to weaken or undermine civilian control remains a matter of debate and speculation. However, for the optimists, the New Japan seems very unlikely to make the same mistakes as the pre-war elite. The world, the Japanese role in this new world and indeed Japan itself have all changed, rendering overly pessimistic forecasts of the return of militarism overstated if not completely misguided. Japan's existing system of civilian control still remains a formidable edifice which will ensure political supremacy, and indeed may offer the new nations of communist Europe a model to control their nascent military establishments in the new market-driven order.

Notes and References

1. The clearest definition is to be found in his 'Civil-Military Relations', in *International Encyclopedia of the Social Sciences* (New York: Free Press, 1968).
2. Bengt Abrahammson, *Military Professionalism* (New York: Sage Publications, 1972) pp. 153–7.
3. A detailed history of the Meiji Military establishment is Ito Kobun's 'Meiji Kokka Ni Okeru Seigun kankeill' (Civil–Military Relations in the Meiji State), *Boei Ronshu*, vol. 7, no. 2, Nov. 1968, pp. 3–13.
4. The best analysis remains Roger Hackett, 'Yamagata Aritomo' in *The Rise of Modern Japan* (Cambridge, Massachusetts: Harvard University Press, 1971).
5. See my 'The Evolution of a General Staff System in the Imperial Japanese Navy', in *Proceedings of the British Association of Japanese Studies*, vol. 4, 1979, G. Daniels (ed.), pp. 73–86.
6. The definitive study of the Japanese Genro is Lesley Connors, *The Emperor's Advisor: Saionji Kinmochi and Prewar Japanese Politics* (London: Croom Helm, 1987).
7. These concepts are taken from the seminal work of J. Crowley on the military establishment and cited in his 'From Closed Door to Empire: The Formation of the Modern Military Establishment', in B. Silberman and H. Harootunian (eds), *Japan in Crisis: Essays in Taisho Democracy* (Princeton University Press, 1966) pp. 261–87.
8. The most recent detailed study of internal naval arms control and Japanese civil–military relations is I. Gow, 'Kato Kanji and the Politics of Naval

Arms Limitation in the Japanese Navy in Politics and Politics in the Japanese Navy', PhD, Sheffield University, 1984.

9. However Cabinet interpretations did permit the PM to occupy the position of Service minister concurrently and pro tem. This happened during the Washington Conference (1921–2) and London Conference (1930) when the Navy Minister attended as chief plenipotentiary. The army was adamant, however, that it would never permit this 'infringement' of the Supreme Command.

10. This interpretation has its basis in the work of Yale Candee Maxon, *Control of Japanese Foreign Policy* (University of California, 1957), which is a work over-reliant on the documents of the Imperial Military Tribunal, Tokyo's prosecution documents.

11. See Asada Sadao, 'Japanese Admirals and the Politics of Naval Arms Limitation', in Gerald Jordan (ed.), *Naval Warfare in the Twentieth Century* (London: Croom Helm, 1977) for an early analysis of this. See also Gow's *Kato Kanji* . . . .

12. See *Jieitai Junenshi* (Ten-year History of the SDF) (Tokyo, 1961), by Jieitai Junenshi and Henshu Iinkai (eds).

13. See my comments in J. Chapman, I. Gow and R. Drifte, *Japan's Quest for Comprehensive Security* (London: Pinter, 1983) pp. 71–6.

14. The House of Representatives established a special Committee on Security on 1 April 1980, and this was followed by a House of Councillors' Special Committee on security in January 1981. The Ministerial Council on security problems within the Cabinet was set up in December 1980.

15. The Chief of the combined Chiefs of Staff (*Bakuyro Kaigi-Cho*) is already an official appointed by Imperial Attestation (i.e. by attending the Palace). The best recent analysis of civilian control issues is Hirose Katsuya, *Kanryo to Gunjin: Bunmin Tosei no Genkai* (Bureaucrats and Military Men: The Limits of Civilian Control) (Tokyo: Iwanami Shoten, 1989).

16. For a completely different view of participation in the UN, citing the Japanese right to be a conscientious objector nation within the UN, see Makoto Oda, *Gekkan Asahi*, January 1991, *A Nation of Conscientious Objectors* (Ryôshinteki hansenka no kuni).

17. See *Tokyo Shinbun*, 4 October 1991, for the discussion of the five Principals and their negotiations in the legislation.

4 Japan's Changing Defence Policy
Tomohisa Sakanaka

I INTRODUCTION

The end of the Cold War in the West has opened the way for a relaxation of tensions in the West Pacific. Although the military stand-off on the Korean Peninsula continues, the establishment of diplomatic relations between South Korea and the Soviet Union in September 1990 helped open the way for simultaneous admission into the United Nations of both the Republic of Korea and the Korean Democratic Republic. Progress is also being made in Indochina with acceptance of the United Nations' (UN) plan for restoring peace in Cambodia. The break-up of the Soviet Union and the inauguration of the Commonwealth of Independent States (CIS) also presents the possibility of re-directing Japanese defence policy. Just as Japan's defence capabilities were built up in a climate of East–West confrontation, the end of that confrontation should have a significant impact. An accelerated modernisation of Japanese defence capabilities and greater US–Japanese cooperation were a direct response to the expansion of Soviet Far East Army in the second half of the 1970s. But with the dissolution of the Soviet Union the Japanese government in December 1991 began a reconsideration of its new Mid-Term Defense Program (fiscal years 1991–5) and to examine the question of reorganising defence capabilities and cutting procurement of major equipment.

II EAST ASIA AND THE END OF THE COLD WAR

The wave of liberation which first swelled in Hungary in May 1989, and later Poland, rapidly swept across Eastern Europe and led to the fall of one communist government after another. On 9 November of the same year, the Berlin Wall, the powerful symbol of the East–West divide, came down and allowed for German reunification on 3 October 1990. Although the Soviet military

69

continued to occupy countries of Eastern Europe, it lost both the ability and the will to control those countries through military force. The Soviet Union itself abolished the monopoly of the Communist Party, promoted liberalisation and democratisation and adopted policies aimed at economic development through the introduction of a market economy. All of this has led to a situation of 'chaos' in the former superpower, making it difficult to predict what the future may bring.

In the 1970s and 1980s the 'Soviet threat' was one of military aggression, but the dissolution of the 'Soviet empire' in the early 1990s has led to a vigorous debate concerning the concept of 'Soviet threat'. One issue in the debate has been over the appropriate response if the CIS were to use military force to put down the Republics' demands for separation and independence. In addition, there is discussion of the serious problems arising from nuclear weapons spreading throughout the Republics, and the possibility of Soviet nuclear experts working in the Third World.

Changes in the Soviet Union have also brought new developments to the security situation in East Asia. The first was a change in Soviet Far Eastern strategy. Former President Gorbachev announced in December 1988 that he would reduce the Soviet military by 500 000 troops, including approximately 20 per cent or 120 000 troops from the Far East Army. His visit to China in May of the following year led to the normalisation of relations between the two countries. Three out of four occupying troops in Mongolia were withdrawn, and the withdrawal of troops from the Sino-Soviet border is under consideration. China appears to have also decided to improve relations with its rival.

The second development occurred on the Korean Peninsula, where the interests of the three nuclear powers are closely intertwined. With South Korea establishing relations with the Soviet Union, and working to strengthen its trade relationship with China after an agreement to establish trade representative offices was concluded, North Korea began to negotiate the establishment of diplomatic relations with Japan in January 1991. These changes present an opportunity to tear down Cold War structures in the Far East and build a new peace and stability.

But there is also a dark side to the positive developments brought on by the end of the Cold War. There is now a greater risk of instability which must be sufficiently considered.

Although the former Soviet Union has begun to reduce its military forces, its military power in East Asia remains as great as ever, and the confrontation on the Korean Peninsula continues. If one looks at the security of Northeast Asia and closely connected Southeast Asia, it is clear that both political and economic instability persists.

The balance of power in the Asia-Pacific region resembles a mosaic. Unlike the bipolar confrontation in Europe, conflict and confrontation in East Asia have a variety of causes whose characteristics change by region and country. The stage of political, economic and social development also varies greatly. At present it is not possible for the region to define a set of common values. Many countries have a history of colonisation and a strong sense of nationalism stemming from that experience. All of this contributes to a fluidity of relations, both friendly and hostile, between nations in the region.

Movement towards an easing of East–West tension in East Asia has been far slower than in Europe where there is now consultation on regional peace and security through the Conference on Security and Cooperation in Europe (CSCE) in which the United States, Canada and 34 European nations participate. There was also agreement in the Conventional Forces in Europe (CFE) negotiations on wide-reaching reductions in five major weapon systems from the Atlantic to the Urals. In the Far East there is neither a forum for regional consultations nor a foundation which would make arms control negotiations possible. Thus, given the opportunities and risks brought on by the end of the Cold War, the time has come to begin the search for a new order in East Asia.

III   CAUSES OF EAST ASIAN INSTABILITY

As suggested above, recent developments have created factors for continued military instability in the region. One cause for instability is the lack of change in the former Soviet Union's military build-up in the West Pacific: what United States and Japanese defence officials see as a continuation of qualitative improvements despite the 'end' of the Cold War. In May 1989 former Soviet Defence Minister Yazov revealed that there were 597 600 troops east of the Urals organised into two groups, one to defend against US and Japanese forces in the Pacific area (326 200 troops) and the other deployed along the Sino-Soviet border (271 400). These are said to maintain 1690 fighter aircraft, 12 600 tanks, 14 300 APC, 16 400 artillery pieces, 55 surface ships and 48 nuclear submarines.[1]

There have been no recent statements regarding Far Eastern deployment of former Soviet troops. Although the CIS has reduced its troop strength aimed at China, a qualitative improvement of the first group continues. The US Defense Department report *Soviet Military Power 1990* expressed the view that future Soviet military reductions will mainly be carried out along the Sino-Soviet border. Reductions of air and ground forces in the Far East Military District aimed at Japan will not be given priority.[2]

The US Defense Department reported in September 1991 in *Military Power in Transition* that the CIS had begun an overall reorganisation of the military and stated that with reductions in its military power there is no longer a realistic possibility that the Soviet Union will mount a sustainable conventional attack against NATO without obviously mobilising over a long period.[3] It is important to recognise that the forces withdrawn under CFE have been redeployed to the Far East region. The same report indicates that since June 1991 16 000 tanks, at least 16 000 APCs and 22 000 artillery pieces have been moved east of the Urals.[4] Thus the counterbalance to disarmament in Europe has been the modernisation of equipment in the Far East.

A second cause for instability is the danger of nuclear proliferation and a subsequent arms race on the Korean Peninsula. Although both North and South call for arms reductions, mutual distrust remains deep. The modernisation and build-up of weapons continues as armies of over one million troops confront each other across the demilitarised zone. Since the Soviet Union established relations with South Korea, the relationship between the Soviet Union, then the CIS, and North Korea has been cool. With the rapid progress in the democratisation of the CIS, North Korea's sense of isolation has probably increased. Achieving peace and stability on the Peninsula is of the utmost urgency for stability in the West Pacific.

The ability of North Korea to possess nuclear weapons is of particular concern. North Korea is said to be operating a research reactor provided by the Soviet Union. US intelligence agencies have detected the construction of a second, larger reactor thought to be capable of producing enough plutonium for about one nuclear weapon annually.[5] At the South–North talks in December 1991 the two nations agreed on the Joint Declaration for Denuclearisation of the Korean Peninsula. Both countries promised not to test, produce, manufacture, introduce, possess, store, deploy or use nuclear weapons.[6] However, complications in the realisation of the Joint Declaration are expected. Even if the North has not yet acquired nuclear weapons, there is no doubt that it has the latent ability to do so. Were it to possess nuclear weapons the impact on the region would be great. South Korea would take countermeasures and there would certainly be an intensification of the nuclear arms race. It would also have a great impact on Japan and China.

A third cause of instability is the impact the Cold War might have on alliance relationships. In the United States the call, particularly in Congress, for a peace dividend is growing louder. According to Defense Secretary Dick Cheney there are plans for deep reductions in military personnel, with the goal of reducing defence expenditures to 4 per cent of GNP by fiscal year 1995. A Department of Defense report, *The Strategic Framework of the Asia-Pacific Region*, released 19 April 1991, revealed a three-stage reduction and reorganisation of the US military deployed in the West Pacific. In the first stage there would be a reduction of 14 000–15 000 personnel from the present 135 000; in the second a reorganisation and reduction of combat units; and further cut-backs in the third.[7]

The United States' East Asian strategy, as far as this report shows, appears to be a gradual reduction of ground forces in units forward deployed in East Asia, placing priority on future naval and air strike capability. This is in line with traditional American thinking since the 1969 Nixon Doctrine was first stated. It is a policy of relying as much as possible on allies for ground forces and supplementing them with American mobile striking power. The importance of the report lies also in the fact that it makes clear that Japanese power projection capability hinders stability. One of the key elements of US strategy is to 'continue to encourage Japan to increase its territorial defence capabilities and enhance its capability to defend its sea lanes . . . while at the same time discouraging any destabilizing development of a power projection capability'.[8]

IV   THE END OF THE COLD WAR AND JAPANESE THREAT PERCEPTION

During the Cold War Japanese defence capability focused primarily on countering the Soviet military threat. Defence officials watched the Soviet Union continuously modernise and build up its military power in the Far East region from the middle of the 1960s. They indicated that 'not only did the build-up intensify the international military situation, but it greatly increased the potential threat to Japan'.[9] The 1990 White Paper on Defense stated that although there was a quantitative reduction in the Soviet Far East military since President Gorbachev announced unilateral reductions in Soviet military forces in the region, the deployment of modern equipment was continuing at a fast pace. It also noted that there had been considerable progress in reorganisation and rationalisation, in addition to modernisation.[10] Although there was no reference to Soviet military power as a 'potential threat' in the 1990 White Paper, the report paid attention to qualitative improvements in the Far East army and revealed a great deal of wariness towards the Soviet Union. The next Defense White Paper published in July 1991 stated that 'although recent changes in the domestic and international environment of the Soviet Union makes it more difficult than before for it to conduct aggressive actions against another country, the severe military situation surrounding our country [Japan] remains unchanged'.[11]

In evaluating threat it is necessary to consider whether to deal with it in terms of the possibility or probability of the threats being realised. Stressing possibility requires a large capacity to respond to possible incidents. Probability, on the other hand, specifies incidents that have a high likelihood of occurring and thus allows for limiting response capability. This, however, leaves open the danger of weakening the ability to respond to unspecified threats. Because it is impossible to maintain the capacity to deal with the entire spectrum of imaginable threats, it is only realistic for Japan to prepare for probable threats based on assessments of the international situation surrounding the country and estimates of the capability and intentions of possible adversaries.

The threats facing Japan can be categorised into three types: The first is global war; the second regional conflict, such as hostilities on the Korea Peninsula; and the third low-intensity conflicts in the Third World threatening access to markets and materials necessary for Japan's economic development. These three threats should be analysed in terms of possibility and probability. Although the possibility of global war still exists, its probability has become extremely low. With nuclear war, in particular, the state of essential equivalence in nuclear capability between the United States and the CIS means that either side would suffer severe retaliation because of an inability to destroy the other side's retaliation capability even through a first strike. This situation is not likely to change in the future; it is realistic to recognise that nuclear weapons have become 'weapons that cannot be used'.

During the Cold War the foremost problem facing Japanese defence planners was how to respond to a conventional global war scenario. Since the early 1980s the Soviet Union through its continuous military build-up has possessed the capability to conduct simultaneous conventional war on fronts in Europe, the Middle East or the Far East. Geographical factors have led it to stress the European front where it is easy to concentrate troops to wage offensive warfare. Hence, the global war scenario at that time predicted that the Soviet Union would take a position of strategic offense on the European front and a position of strategic defence in the Far East.

It is thus hard to imagine that the CIS would launch an attack aimed at the occupation of Japan. It is more likely that military operations against Japan would be offensives against US bases in order to prevent American actions against the CIS and to challenge US control of the seas. An offensive against Japan would be to guarantee passage through the Tsushima, Tsugaru and Soya Straits, to make this possible by partially occupying Hokkaido, and to blockade sea routes and attack ports and means of transportation to increase pressure on Japan. During the period of East–West confrontation Japanese officials focused on dealing with this kind of scenario for conventional global war.

Japan's emphasis since the mid-1980s on air defence of the sea and air surrounding Japan, sea lane defence, and the prevention of landings on Japanese territories stems from this global war scenario. Although the possibility of global war disappeared with the end of the Cold War, maintenance of this defence capability to deal with limited attacks is necessary as a basic defence function of a modern nation. Maintaining this capability provides a way to ensure against 'backsliding' in the military policies of the CIS.

The second type of threat facing Japan is regional conflict. The Iraqi invasion of Kuwait in August 1990 pointed out how grave a threat regional conflicts pose to global peace and security. Because of the dependence of advanced industrialised nations on energy resources from the Middle East, conflicts in that region are a direct threat to global peace and stability. A recent US Defense Department report points out that the focus of the new American strategy had shifted from repelling a global challenge from the Soviet Union to dealing with threats to major regions such as Europe, Southwest Asia and East Asia.[12] Yet if the US military deployed in East Asia were withdrawn, Japan's ability to deal with regional conflict would be limited. As already mentioned above, because the US is opposed to Japan maintaining power projection capability the United States must continue to maintain its forward deployment in East Asia in order to deal effectively with regional conflict. For this reason it is important for Japan to provide stable bases for the US military.

The stability of the Korean Peninsula is of particular importance for Japan. A large-scale military clash on the Peninsula would have a complicated impact on Japanese politics, economics and society. Militarily, it would present the problem of how to protect Japanese aircraft, transport ships and fishing boats and how to respond to territorial infringements by North Korean aircraft and ships. The capacity to deal appropriately with these threats must be maintained in peacetime.

The third threat facing Japan is the threat of low intensity conflicts in the Third World. Domestic instability and instability between countries is likely to occur in the Third World where complicated territorial, ethnic and religious problems abound and there are great disparities in wealth caused by late economic development. It is easy for tensions to escalate into armed conflict. The situation has been complicated recently with the overlapping of population, resource and refugee problems. In the Third World the repeated outbreak of low-intensity conflicts, such as terrorism, drug trafficking, subversive activities and rebellions, as opposed to organised military operations are likely to continue. Supporting this point an early 1990s US Defense Department report also points out that war between the superpowers had been effectively prevented for forty years, but low-level conflict had not been deterred. It emphasised the importance of measures to prevent such conflicts.[13]

In terms of probability a response to these kinds of regional and low-intensity conflicts is a more important issue than a response to a global war. No matter what the country, it is necessary to have a defence capability that is able to deal appropriately with anticipated threats. Japan requires a defence policy which will enable it to respond flexibly to changes in the international environment.

One way in which the Japanese government has attempted to respond to changes in the world is by creating a framework for Japanese participation in UN peacekeeping activities. Limited by a constitution which proscribes war (see Chapter 2 in the present volume), Japan has until recently pursued a policy of not dispatching the Self-Defense Forces (SDF) abroad for military purposes. Japan must now decide, taking account of international opinion, the nature of the SDF's role abroad.

In the near future it will be necessary to consider two issues in connection with the constitutional limitation. The first is the necessity of supporting US activities to maintain the international order. The question of whether or not it is possible to cooperate in areas such as medical, transportation and supply support in addition to financial cooperation must be examined. The second issue has regard to the nature of Japanese cooperation under the United Nations, particularly in peacekeeping activities, given the growing importance of the UN's role in maintaining peace and resolving conflict.

V   JAPANESE DEFENCE BUILD-UP IN THE POST-COLD WAR PERIOD

Japanese defence capabilities since the First Defense Build-Up Plan (1958–60) have increased within the framework of the United States–Japan Security Treaty system. This first plan focused on building up the Ground SDF in order to fill a gap created by the rapid withdrawal of US ground forces after the Korean War. The second plan (1962–6) set out to strengthen the Maritime and Air SDF to allow for a conventional response to aggression in a regional war. The third (1967–71) and the fourth

(1972–6) followed the same policy, introducing new weapons and modernising the SDF.

In October 1976 the National Defense Council approved the 'National Defense Program Outline' setting the basic policy for the defence build-up after the fourth plan. The Outline stipulated that the defence structure should possess the various functions required for national defence, while retaining balanced organisation and deployment including logistical support. It aimed at building a basic defence capability which would make an effective response to 'small-scale, limited aggression' possible. The 1986–90 Mid-Term Defense Program and the new 1991–5 Program are both based on the Outline.

Japan's current defence capabilities are based on:

1. relying on US nuclear deterrence for countering a nuclear threat;
2. cooperating with the United States in responding to large-scale conventional threats;
3. deterring through its own resources limited, small-scale aggression.

It is assumed that the SDF will be responsible for strategic defence, while the US military takes responsibility for strategic offense in response to large-scale aggression. This division of roles was spelled out in the November 1978 Guidelines for Japan–United States Defense Cooperation.

In the 1980s Japan sought to steadily increase its defence spending and to modernise defence capabilities. As a result it achieved the procurement levels called for in the Mid-Term Program. The GSDF (Ground SDF) now consists of 13 infantry divisions and two composite brigades, with an authorised personnel of 180 000 (actual personnel, 160 000). The MSDF (maritime SDF) is organised into four flotillas to defend sea lanes, and 10 regional districts to defend major ports and straits. It possesses six submarine flotillas (16 submarines), 62 surface ships (43 of which are missile-carrying), and 170 aircraft for anti-submarine warfare of which 100 are P3Cs. The ASDF maintains 28 air control and warning wings, ten squadrons of 200 F15s, three F4 squadrons, and one Nike and 5 Patriot groups for air defence.

Although 250 000 authorised Personnel for the SDF is a small number, Japanese ground, maritime and air capabilities represent the most modern in East Asia. Ground units equipped with anti-ship missiles and anti-tank helicopters have a considerable capacity to respond to landing invasions. Air defence with 200 F15s and the Patriot missile is of a high calibre. Both are intended to function effectively in joint manoeuvres with the United States in a global war scenario.

The end of the Cold War and the receding possibility of global war has raised the question of how to analyse the SDF force structure. One way is by comparison with the military power of neighbouring countries, a so-called 'static approach'. A second way is via a 'dynamic approach' of examining the SDF in terms of probable threat scenarios.

The problem with using the static approach is that there is no objective index for determining the correct level of defence power. As the threat level each country faces differs and divergent domestic and international conditions exist, there are naturally differences in the character and level of defence capabilities required. Table 4.1 shows that if Japan's defence capability is compared with that of neighbouring and Western countries in terms of population, area, or the ratio of defence spending to GNP, it falls below the international average in all indicators. This is because the threat to Japan even during the Cold War was comparatively low and also reliance on the US military under the United States–Japan security treaty system was great.

Insufficient reserve forces are particularly noticeable. The SDF has only 46 000 reserves for 250 000 active forces. NATO countries maintain reserve forces basically equal to their active forces. If Japan were to maintain reserve forces on a similar scale, it would perhaps be possible for it to greatly reduce its active forces. Yet the political climate in Japan makes a reserve force of that size next to impossible. Thus from a defence policy standpoint it would be extremely difficult to reduce the current number of active personnel.

Another major characteristic of Japanese defence policy is the complementarity of United States–Japanese military power. Japan pursues a policy of strategic defence and reliance on the US offensive force. The GSDF is a garrison type force and does not have counterattack capabilities to conduct amphibious operations. The MSDF is composed of ships and aircraft intended for anti-submarine operations and lacks power projection

Table 4.1 Comparison of 1990 defence strengths

	Pop. (A)	Area (B)	Military personnel (C)	Reserves (D)	C/A (× 100)	C/B	D/C	GNP[1]
Japan	122 265	378	249	46	2.04	0.65	0.18	1.0
South Korea	41 575	99	750	4 500	18.03	7.58	6.00	4.4
China	1 088 570	9 561	3 030	1 200+	2.78	0.32	0.40	1.7
Germany[2]	61 171	249	476	853	7.66	1.88	1.82	2.2
United Kingdom	56 930	244	360	340	5.37	1.25	1.11	3.7
US	243 934	9 373	2 117	1 613	8.68	0.22	0.76	5.4
USSR	283 100	22 402	3 988	5 602	14.12	0.18	1.40	11.1

Notes:
[1] Defence spending as percentage of GNP.
[2] Figures are for West Germany.
Population, active and reserve forces, area by 1000 square kilometres.
Also: C/A = Military personnel per capita

C/B = Military personnel to territory

D/C = Active military personnel to reserves

Source: *Military Balance 1991–1992*.

capability similar to the US carrier task forces. The ASDF (Air SDF) is centred around air defence and does not have offensive capabilities for attacking an adversary's bases. Under the policy of 'exclusive defence' the SDF relies on the United States for offensive power. This reliance contains the potential to restrict Japan's defence capability if US support diminishes. It is thus necessary to examine the question of how to fill in the gaps in Japan's defence functions that would arise if the United States were to withdraw from East Asia.

In examining the SDF from a dynamic standpoint one must consider that Japanese defence capability was built up from the early 1980s under the global war scenario. It was structured as part of a system of Soviet containment in keeping with the global expansion of the Soviet threat. Although the 1976 'Defense Program Outline' which called for the build-up of defence functions necessary to a modern nation did not indicate a 'specific threat', in actuality it placed priority on responding in cooperation with the United States to the growing Soviet threat. This kind of defence capability is still effective against large-scale regional conflict even after the Cold War. Yet this ties defence build-up to cooperation with American operations.

For example, air defence, the defence of sea lanes and defending against landings placed priority on joint United States–Japan manoeuvres in order to counter the global threat of the former Soviet Union. But since Japan must also be able to respond on its own, this has led to a lack of flexibility for dealing with regional and small-scale conflicts. While it is necessary to expand United States–Japan defence cooperation, being the cornerstone of Japan's defence policy, it is also necessary to develop the ability for flexible responses to limited conflict.

A second problem concerns how to respond to conflict in the Third World. Japan's defence capabilities focus on Japanese defence of its land, sea and air. In the Cold War this kind of territorial defence was structured as part of the containment of the communist bloc. With the end of that 'war' it has become necessary from an international standpoint to reconsider the role of defence capabilities. NATO, for example, has reorganised its forces and created different types of multinational contingency forces. It has begun to examine the question of how to deal with regional instability. As mentioned above, the area of international cooperation in defence-related issues has also become important for Japan. It needs to develop countermeasures based around the United States–Japan treaty system to deal with Third World conflict and to participate in UN peacekeeping activities.

VI CONCLUSION

The change in the strategic environment which the end of the Cold War represents has made it necessary to re-examine all aspects of Japanese security policy and to establish a comprehensive long-term strategy. A number of issues must be taken into consideration in formulating this kind of strategy.

The first is that as Japan approaches the twenty-first century it must decide how to view its responsibilities in a world of growing interdependence and 'globalised' security. Even if Japan were to seek to limit its role to economic areas, as the world's second largest economic power its policies will have a considerable impact on global peace and security. It will become impossible to avoid political decisions about the kind of role Japan should play with regard to security issues. From now on Japan's perception of its role will become increasingly important.

The second issue is the extreme importance of the US–Japanese relationship as we head into the twenty-fist century. Although the United States–Japan Security Treaty system's role as part of an anti-communist strategy disappeared with the end of the Cold War, its importance as a system of stability for both East Asia and the world at large remains as important as ever. For the foreseeable future there is no country which could assume America's security role in East Asia. A 'division of labour' between the United States and Japan will also be necessary in the future. This will not be a division of costs to maintain the treaty system based on a short-term perspective, but must be a division of labour resting firmly on a long-term perspective of designing a system of international order for the twenty-first century.

Thirdly, while regional cooperation in East Asia is important, realising this goal will not be easy because of differences in political systems, disparities in economic and social development, and historical factors. Even so, a dialogue on East Asian peace and security is important, and the end of the Cold War has offered an important opportunity for building a framework for stability in the Asia-Pacific region. The need is now important to look towards developing a dialogue on regional cooperation, confidence-building measures, regional disarmament and regional security. The end of the Cold War has provided Japan with a golden opportunity. During the Cold War the effort to counter the Soviet threat was of grave importance, but now there is a chance to re-examine and realise a new role for Japan in the international community.

Notes and References

1. *Asahi Shimbun*, 29 May 1989.
2. US Department of Defense, *The Soviet Military Power 1990*, September 1990, pp. 97–9.
3. US Department of Defense, *Military Power in Transition*, September 1991, p. 11.

4. Ibid, p. 12.

5. *Arms Control Today*, March 1991, p. 9.

6. *Asahi Shimbun*, 1 January 1991.

7. US Department of Defense, *The Strategic Framework for the Asian Pacific Rim: Looking Towards the 21st Century*, April 1991, pp. 13–14.

8. Ibid, p. 18.

9. Japanese Defense Agency, *Boeihakusho* (Defense White Paper), August 1986, Section 27.

10. Japanese Defense Agency, *Boeihakusho*, September 1990, Section 46.

11. Japanese Defense Agency, *Boeihakusho*, September 1991, Section 45–6.

12. Report of the Secretary of Defense to the President and the Congress, January 1990, p. 5.

13. Ibid, p. 6.

5 Japan and Regional Security

S. Javed Maswood

I INTRODUCTION

In recent years there has been considerable analysis and speculation, both within and outside Japan, on the future direction of
Japanese foreign policy. Although this is not an altogether new
preoccupation, there does appear to be a greater willingness to
discuss openly such previously 'taboo' subjects like constitutional
reform and the possible uses of Japanese military in international
peacekeeping roles. The Japanese Constitution has been an important determinant of foreign and defence policies. Article 9 of
the Constitution and the spirit of international pacifism gave
Japan a unique position within the international community. It
led to a relative withdrawal and disengagement from international political issues. These precepts of Japanese foreign policy
are being increasingly questioned. For example, according to
one Japanese analyst it is time for Japan to take a second look at
the constitutional arrangements even if the review resulted in
reaffirming existing arrangements.[1] The liberalisation of the parameters of debate, the present soul-searching and impetus to
redefine Japan's international role and security interests can be
attributed to the changed international conditions like, for example, the disintegration of the Soviet threat and end of the Cold
War.

In the past, discussions of Japan's international role usually
accompanied an unqualified disclaimer that Japan would seek
neither military power status nor a military role. This was
necessary not only to placate domestic opposition groups but
equally to reassure other Asian countries that Japan would not
pose a threat to their security. The Fukuda Doctrine explicated
Japan's peaceful intentions most clearly. Recent statements by
senior Japanese government officials, however, suggest a new
thinking. The Japanese Ambassador to Thailand, Okazaki Hisa-
hiko, explained that it was impossible for anyone to give un-
equivocal and firm guarantees against Japan becoming a military

power. The only guarantee that could be given against that possibility was simply that it was not in Japan's economic interests to tread the path of military power.[2] While there is no reason to expect Japan to begin flexing its military muscle, it is still necessary to try and fathom the depth of Japanese willingness to be more active and the direction this activist foreign policy might take. In this chapter, the focus will largely be on Japanese relations with the countries of Southeast Asia and the evolving structure of regional security.

II REGIONALISM AND GLOBALISM IN JAPANESE FOREIGN POLICY

It is hard to envisage Japan reverting to prewar patterns of militarism, even though the Diet, in mid-June 1992, finally approved new legislation to end the longstanding proscription on the dispatch of Japanese military personnel overseas. The legislation will allow Japanese Self Defense Forces to participate in United Nations peacekeeping operations, although in a non-military capacity. The Japanese government had pushed hard for such legislation following the Gulf crisis and Western criticism that Japan's contribution to the Multinational Task Force was inadequate, even if financially generous. These attempts and the subsequent passage of the PKO (peacekeeping operations) legislation has heightened concern in some of the regional countries, most notably South Korea, that it would open the floodgates to a military role for Japan on the international stage and that Japan would seek to complement its economic power with military power. These fears are reinforced by the fact that Japanese defence spendings, at around 1 per cent of GNP, are already the third largest in the world and more than the combined military spending of all Asian countries if the Peoples Republic of China is excluded. Nonetheless, as the Japanese government argues, the force posture of the Japanese military establishment is geared for defensive operations rather than for offense and power projection.

The Japanese government, although sensitive to regional concerns appears determined not to let these negative reactions dictate foreign policy directions. The Japanese Prime Minister Miyazawa Kiichi, in his first ever address to the Korean

Parliament in January 1992, assured the 230 parliamentarians that while Japan did not seek a military role, the government was resolute in its commitment to participate in future peacekeeping operations of the United Nations.[3] The forthrightness with which Miyazawa presented the government's position was exceptional, given the Japanese government's reputation for avoiding controversy. The boldness can be attributed, at least partially, to the fact that resistance, within and outside Japan, to a larger Japanese role, including a military role, is weaker than in the past. The regional countries are less united and less steeped in their mistrust of Japan. For example, the Prime Minister of Thailand, in an interview in September 1991 stated that there was no reason to reject participation by the Japanese military in United Nations' operations.[4] This, and similar sentiments expressed by governments in Singapore and Australia, suggests that misgivings about Japan are not as widely shared as in the past. It is instructive also that the visit to Southeast Asia in 1991 by the Japanese Emperor did not incite the same passion and protest activity that had marred the visit of Prime Minister Tanaka in 1975.

There is, therefore, a greater opportunity for Japan to become politically active on the international stage. Similarly, it is certain that increasing international financial obligations on the Japanese government will push it to seek influence commensurate to its financial commitments. In 1992–3 Japan became the second biggest source of funds for the United Nations and Japan was also extremely generous in contributing to the costs associated with the operations of the multilateral task force during the Gulf crisis. It is not very likely that the Japanese government will resign itself to 'taxation without representation'. There is no suggestion, however, that Japan is poised to acquire military status or to play a military role in the region.

In view of the above, interest in the future direction of Japanese foreign policy is understandable. Another reason for this interest is the series of events that has altered the political map of the world and with it the political and security environment confronting Japan. Paradoxically for Japan, the end of the Cold War is both a welcome and a threatening development. On the one hand, the 'new world order' is not dominated by threat perceptions and presumably less prone to conflict but, on the other hand, there are considerable uncertainties associated with the changing patterns of international relations. The latter aspect of the new world order is significant because Japan's economic prosperity depends on stability and predictability of international and trade relations. With a higher potential for instability, the Japanese government can ill-afford to remain a rule taker within the international system and we can expect it to become more involved in ensuring conditions conducive to its economic interests.

Although Japanese economic interests are global, its political and security interests are in safeguarding regional peace and stability. As such, a larger political role for the Japanese government will in the first instance be regional in scope. This is consistent with and highlighted in the recent declaration on global partnership that was issued following the visit to Japan by US President Bush in January 1992.[5] The idea of a global partnership reflects the fact that Japan and the United States together account for 40 per cent of global GNP and that their bilateral relationship and problems have global consequences. Within the context of a global partnership Japan can be expected to acquire the status of a regional political power. In terms of economic influence, of course, Japan already is a global power. Thus, even though the Asian countries are increasingly becoming more economically important for Japan, as Inoguchi Takashi argues, 'Asia' as an economic unit is too small for Japan and that it would be detrimental for Japan to become an 'Asian' power. Inoguchi further argues that Asia is small also for the continued growth and prosperity of other Asian countries, now that many have adopted a growth strategy based on export-oriented industrialisation rather than import substitution. The success of the former depends on access to foreign markets and this underscores the importance of reaching beyond Asia to other large markets, like the United States. Inoguchi suggests that the main task for the Japanese government is to preempt the possibility of narrow regionalism and the formation of a trading bloc.[6] From this point of view, even though Japan's political role is likely to be regional in nature, Japanese interests are in pursuing open, rather than closed, regionalism.

Within the region, the Association of Southeast Asian Nations (ASEAN) has, since its establishment in 1967, emerged as the leading regional institution. ASEAN is composed of six of the Southeast Asian countries, Indonesia, Malaysia, Singapore,

Philippines, Thailand and Brunei. The remaining four South-east Asian states, Vietnam, Cambodia, Laos and Myanmar, have also indicated their interest in joining ASEAN and it is expected these countries will be granted membership in the not too distant future. Although ASEAN is not yet a truly regional institution, it has acquired considerable influence by presenting a united front to the rest of the world. Given the important position of ASEAN in regional affairs, a larger Japanese regional role is likely to be based on a constructive partnership with ASEAN.

Even though it is less mistrustful of Japan, ASEAN is mindful of the danger that it could be swamped by a politically powerful and active Japan. From its perspective, therefore, the least problematic of all future scenarios would be for Japan to limit itself to an economic role. This would be desirable not only because of the potential benefits to be had from a further deepening of regional division of labour and increases in Japanese investments but also because it would lay the basis of a relationship based on trust. The Prime Minister of Singapore, Goh Chok Tong, in his keynote address to the Pacific Economic Cooperation Conference in late May 1991 stated that:

> Japan can win over the trust of its neighbours by demonstrating to them the same magnanimity of spirit that the United States once displayed towards Japan. Japan should make major investments in the Asia-Pacific region. It should transfer technology, skills and management to its Asia-Pacific neighbours.[7]

Under these circumstances any attempt by Japan to introduce a non-economic dimension to its regional policies will generate some measure of controversy and resistance. More than forty-five years after the Second World War, there is still some lingering suspicion that Japan may not have fully forsaken its militarist past and that democratic values may not have displaced earlier patterns of authoritarianism. It is understandable, therefore, that Japanese remilitarisation in the postwar period has been the biggest source of concern. Certainly, the strength of these concerns has weakened over time but there is, as yet, no ready acceptance of Japan by all the countries of Southeast Asia. Most regional countries do, however, acknowledge the inevitability of a much larger Japanese role given its economic dominance. Japan is not only emerging as the linchpin of an Asian division of labour but also the focus of Japanese direct foreign investment (DFI) is shifting to Asia and ASEAN. Between 1950 and 1990 the whole of Asia received only 15 per cent of Japanese foreign direct investment but a survey of 115 Japanese companies revealed that between 1992 and 1994 roughly half of their DFI will be in the Asian countries, including about 25 per cent in ASEAN.[8] In terms of trade patterns, the export dependence of ASEAN on Japan is relatively high. In 1989, it ranged from 8.5 per cent for Singapore, 20.1 per cent for the Philippines, 16.9 per cent for Malaysia, and a high of 40 per cent for Indonesia.[9] The structure of Japanese economic relations with the Asian countries, thus, will inevitably translate to increased influence for Japan. What is of concern to the regional countries is the context within which Japan acquires and executes its role.

The two scenarios are for Japan to become an autonomous and independent player, on the one hand, or to remain within the parametric constraints of the United States–Japan security alliance, on the other. If there is apprehension, it is of a Japan that is both significantly militarised and also an independent player in regional affairs. As it is, there already are misgivings about Japan's present state of militarisation. The latest Korean Defense White Paper, released in late October 1991, pointed to Japan as the security threat which had, contrary to pretensions, become a force with offensive capabilities.[10]

From the perspective of the regional countries, the United States–Japan security treaty is looked upon as providing assurances that the United States will have a restraining influence on Japan. This view is shared by most of the regional countries, and not only in Southeast Asia. It follows from this that most of the regional countries prefer continued American involvement in the region. The familiarity that has been the result of over fifty years of American involvement in the region is preferred to the uncertainties of some other country, including Japan's, expanded and unchecked role. The fear is that in the post Cold War environment, the United States might be inclined to pare down its presence to the extent of creating a power vacuum which other countries might be tempted to fill. These concerns were rekindled in late 1991 when President Bush cancelled a scheduled visit to the region. It seemed to suggest that Asia was not a high priority region on the American agenda. Although the visit was hastily rescheduled, the damage had been done and President Bush was forced to reiterate and reassure regional leaders throughout his Australian and Asian tour of American commitment to the region. During his stay in Singapore, the American president delivered the annual Singapore Lecture and stated that 'we will maintain a visible, credible presence in the Asia-Pacific region with our forward deployed forces, and through bilateral defence arrangements with nations of the region'.[11]

American presence in the region is reassuring also because of the view that as long as Japan and the United States maintain their alliance relationship, Japan will not be tempted to undertake a significant expansion of its defence capabilities. Even the United States recognises the importance of its security treaty with Japan. The Commander of the US marine corps in Japan, General Henry C. Stackpole III, stated that:

> Japan will beef up what is already a very, very potent military force if US forces withdraw. No one wants a rearmed, resurgent Japan. So we are a cap in the bottle, if you will.[12]

If the United States withdrew from the region there is the additional danger that Japan might develop nuclear weapons capability if, as is reported, North Korea is successful in developing a nuclear potential. The United States–Japan Security Treaty extends a nuclear umbrella over Japan but if this is withdrawn there may be a strong incentive for Japan to renege on its non-nuclear principles. As long as Japan remains under the US nuclear umbrella there is little danger of large-scale Japanese militarisation or a decision to acquire nuclear capabilities.

Within Japan, although there are solitary voices that question the validity of the security treaty in the post Cold War period, there is no concerted push to revise the essentials of the relationship with the United States. But the relationship will continue to evolve and its importance will be defined not simply in terms of a bilateral agreement but rather as an important aspect of an emerging multilateral framework.

The American presence in the region in the postwar period was part of a global strategy to contain the spread of Soviet influence and communism but this threat has since disappeared. Yet for reasons mentioned above, American presence continues to be regarded as important by most of the countries in the region. Only Malaysia appears to be mildly ambivalent about American presence. The Malaysian government in December 1990 proposed the formation of an East Asian Economic Group, which would exclude 'outside' powers, including the United States, Australia and Canada. Later, the proposal was changed to an East Asian Economic Caucus to dissociate it from suggestions that it was intended to become a regional trading bloc. The United States has been strongly critical of it as a departure from the principles of global free trade, despite its own sponsorship of a North American Free Trade Agreement. The American government is more concerned with its own exclusion from the concept, and for this same reason the proposal has not attracted much regional support. Most of the East Asian countries, including Japan, continue to see American presence as desirable from the point of regional stability. If Malaysia has been at the forefront of attempts to form a narrowly defined regional grouping, it, together with Thailand, is also more receptive than others of a larger regional role for Japan. In its economic policy outlook, Malaysia has emphasised the importance of learning from Japan, a policy orientation encapsulated in exhortations to 'Look East'. Malaysia also sees Japan as a bulwark against possible Chinese hegemony.[13]

III TOWARD A MULTILATERAL REGIONAL SECURITY SYSTEM

The symbols of American commitment to the region were the large Clark Air Base and the Subic Naval Base in the Philippines. The latter, however, has already been closed down and the United States is expected to vacate Clark Air Base in 1993. Although no other Southeast Asian country is in a position to offer alternative and comparable basing arrangements, Singapore, in early 1992, formalised an agreement with the United States to permit a logistics base on the island plus ship repair and maintenance facilities. There was some initial discontent within ASEAN that the decision of the Singapore government contravened commitments to preserve a Zone of Peace, Freedom and Neutrality (Zopfan) but it was subsequently reported that ASEAN officials were in fundamental agreement that the presence of American forces in the region was 'a positive and stabilizing factor'.[14]

For Japan, as well, the security relationship with the United States continues to be important, even though the 1991 Defense White Paper formally acknowledged that the Soviet Union was no longer a threat to Japan and that Japan's security environment had, consequently, improved. Since then, the dissolution of the Soviet Union has further improved the security outlook for Japan. Japan's security relationship with the United States has been the bedrock of its foreign policy and retains, even today, a significance undiminished by changes in threat perceptions. It complements the very important economic relationship between the two countries and neither side is particularly keen to have it fall by the wayside. At the same time, however, the United States–Japan bilateral security relationship will have to adjust to changing international conditions. In future its significance will be largely within a multilateral framework. This does not necessarily mean a formalised multilateral security arrangement but rather the creation of a structure that allows for multilateral security concerns to be discussed and resolved. Such a new framework can also become the basis for ensuring American commitment to and presence in the region.

At the July 1991 ASEAN Post Ministerial Conference (PMC), the Japanese Foreign Minister Nakayama Toru proposed expanding the scope of the PMC to include security-related issues. The PMCs, held immediately following ASEAN summit meetings, bring together ASEAN and its dialogue partners which include Australia, Japan, the EC, Canada, South Korea and the United States. Addressing the conference, Foreign Minister Nakayama stated that: 'I believe it would be meaningful and timely to use the ASEAN post-ministerial conference as a process of political discussions, designed to improve the sense of security among us.'[15] This suggested the need for a security community and for redefining security away from earlier preoccupations with external threat perceptions. For much of the postwar period, the concept of security, whether for one country or for a bloc, had been interpreted in competitive terms to mean security against threats emanating from external sources. This zero-sum concept of security is rapidly being eroded and the Japanese proposal emphasised the importance of rethinking security as a positive-sum game. In this reformulation, a security community is defined not by the exclusion of unlike units and interests but is inclusive of dissimilar units, and disparate and conflicting interests. The Japanese proposal contained the hope that an inclusionary security community would be better able to deal with the numerous political and territorial disputes within the region. It suggested that the regional countries explore the possibility of 'sharing' security which might lead to the emergence of a positive-sum security environment that would be beneficial to all the countries.[16] The idea of sharing security borrowed from, but did not propose, a direct transplant of European-style security cooperation in this region.

Coming from Japan the proposal for an extended security forum was unusual. As mentioned, the Japanese government in the past had assiduously avoided security and military issues and preferred to concentrate on less controversial economic matters. And the proposal for a security dialogue was controversial. It was not, however, an entirely original proposal.

A year earlier in July 1990 the Australian Foreign Minister, Gareth Evans, had proposed an Asian version of the Conference on Security and Cooperation in Europe (CSCE) as a means for promoting common regional security. He argued that:

> it is time to think in terms of a cooperative rather than a confrontational approach to regional security. We simply cannot act as if we were in a time warp, with all the verities of the Cold War period still applicable.[17]

The US reaction to this was swift and hostile, and provoked an angry exchange of letters between Gareth Evans and American Secretary of State, James Baker. The US rejection of the Australian proposal was based on the view that bilateral arrangements, with Japan, Australia, the Philippines, South Korea and Thailand, had successfully underpinned regional security in the past and that there was no need to tamper with a working system, especially as the Asian security environment was different from that in Europe. The American government emphasised that unlike in Europe, the Asia-Pacific region was characterised by numerous divisions which made the European model unworkable in the Asian context. The United States also expressed fears that regional security proposals provided the Soviet Union with a platform for demanding naval disarmament which would be more disadvantageous to a naval power like the United States than to the Soviet Union. In October 1991, Gareth Evans defended his proposal saying there was no suggestion of a simplistic transfer of European models, and observed, with a touch of irony, that what had initially been dismissed as too radical had now come to be 'regarded as boringly commonplace'.[18]

Understandably the Australian government was quick to support the Japanese proposal and Evans expressed satisfaction that the idea had been placed on the ASEAN agenda. More significantly the US government offered guarded support. It could not, of course, dismiss a proposal from the Japanese government as easily as it had done the Australian proposal. The reaction was not enthusiastic but nor did the US government reject it. Secretary of State James Baker commented simply that 'I think it is something that should be considered but beyond that I am not prepared to go.'[19] Later, he had warmed to the proposal considerably more. In a speech delivered in Japan in late 1991, he expressed his conviction that increased multilateral policy initiatives will further strengthen the bilateral security relationship between Japan and the United States.[20] The shift in the American position can be attributed to several factors. It may have resigned itself to the inevitability of multilateral regional arrangements but, equally, the United States was no longer concerned to the same extent with the strategic equation between itself and the Soviet Union. The Cold War was rapidly being wound down, with 'victory' to the United States. The American victory meant that US foreign policy no longer had to be tempered by the Cold War considerations of the past. It also made a difference that the Japanese proposal accepted existing ASEAN structures as the basis of regional security dialogue rather than giving the impression of wanting to emulate the CSCE and establish new institutions.

In so far as the proposal called for extending the scope of existing ASEAN institutions it was clear that progress would be predicated on the response of the ASEAN countries themselves. ASEAN, of course, does not always speak with a single voice. There have been earlier suggestions from individual ASEAN leaders that the regional body should upgrade existing levels of security cooperation but these have not received unanimous support of the members. As an institution ASEAN has not favoured deepening of security and defence relations even among themselves and preferred, instead, to rely on bilateral arrangements, where necessary. In the past, Malaysia had floated the idea of a 'defence community' and a 'joint command'; Indonesia the idea of a 'joint defence council'; and Thailand and Philippines had advocated multilateral military exercises, but, as Acharya points out, none of these proposals received broad support from ASEAN decision-makers at the time.[21]

Given that ASEAN has been unable to agree on intra-ASEAN security arrangements, it is not surprising that the immediate reaction to the Japanese proposal was less than enthusiastic. The Indonesian Foreign Minister, Ali Alatas, questioned the need for formal restructuring of existing arrangements as security-related issues were not precluded from discussion at the PMC. Indeed, the initiation of PMC and dialogue between ASEAN and other large actors in 1979 was not unrelated to threat perception. It followed the Vietnamese invasion of Cambodia in 1978 and the fear that Vietnam would emerge as the regional hegemon. This act of aggression was strongly denounced by the ASEAN countries and it also forced them to seek external friends and allies. The 1979 PMC was attended by the United States, Japan, Australia, New Zealand and the European Community and, as Michael Antolik observes, ASEAN members were glad that the United States had decided to attend 'to restore some sort of strategic balance in the region'.[22] Confronted with the Japanese proposal, there was hesitation about formalising security arrangements but instead of rejecting it outright the Indonesian Foreign Minister commented simply that 'We would want to think more deeply . . . about institutionalising it [PMC] so quickly.'[23]

Apart from intra-ASEAN disagreement as to the nature of desirable security cooperation within the group, ASEAN has had particular misgivings about broadening the scope of such arrangements. In the past, it sought to limit outside involvement in the region as a way of ensuring its own independence from domination by external powers. As such, the Japanese proposal may have been interpreted as attempts by Japan to dominate ASEAN and extend its military influence in the region. Military domination, however, was clearly not a factor behind the Japanese proposal and, as the government pointed out, the Japanese government had only assumed responsibility for sea lane defence extending to 1000 nautical miles. It argued, there-fore, that the Southeast Asian countries had nothing to fear from Japan.

It was obvious that the Japanese proposal had to overcome the inertia and misgivings of the ASEAN member countries before it could be realised. Within ASEAN there were two divergent views on the future security role of the association. On the one hand, the governments of Singapore, Thailand and Philippines voiced the opinion that a fresh approach was required and questioned the continued relevance of Zopfan.

On the other hand, the Indonesian and the Malaysian govern-ments wanted a commitment to the original principles of neutrality.[24] Nevertheless, six months later, at the fourth ASEAN summit meeting held in Singapore in late January 1992 the leaders of the six countries agreed to the broad thrust of regional security proposals. In keeping with the tradition of accommodating intra-ASEAN diversity the agreement was based on balancing the conflicting views. The Singapore declara-tion at the end of the summit stated that the PMC would become a forum for extended dialogues in political and security matters. According to one ASEAN official: 'The PMC has been evolving as a forum for dialogue on security; now it is clearly stated.'[25] The Declaration also stated that the members would continue efforts to realise Zopfan.

There were several factors that made agreement possible. Most important, of course, was the rapid transformation of the international geopolitical situation and a growing perception that institutional mechanisms were necessary to deal with future uncertainties. Moreover, the idea of an extended security dia-logue that would enhance the security of all participants was not too dissimilar to the original ideas that had spawned ASEAN. Although the 'shared' approach to security may have become globally relevant only in the mid-1980s, these principles were the driving force behind the formation of ASEAN. It was the ASEAN members' concern with internal threats to regime secur-ity and survival that led to the realisation that they could each enhance their security without increasing the insecurity of others through mechanisms to ensure cooperation. This shared approach to security driven by domestic concerns and threats became diluted, though not entirely lost, with the emergence of the Vietnamese threat in the late 1970s following its invasion of

Cambodia. This galvanised the association to become more mindful of external threats to security. The return to the original principles of a shared security community, in turn, was made possible by the Cambodian peace agreement of October 1991 and the promise of peace in Indochina.

It is necessary now to consider the implications of this agreement to institute a broad-based security dialogue. Foremost among the implications is that it will further erode the barriers that have prevented Japanese reintegration into the region. Following its defeat in the Second World War Japan had not only pulled out of the region as part of the conditions imposed by Western occupation authorities but Japan had also been pushed out by the hostility directed towards it by people who had suffered under Japanese occupation and colonisation. Unlike West Germany which was rehabilitated into a transformed West European system, Japan was forced to look beyond its immediate neighbourhood for economic and security links. The emerging regional security structure enables Japan to participate fully and equally with other countries of the Asia-Pacific in discussions of security and political issues. This will not, however, detract from the present utility of the security alliance with the United States but will instead be a supplement to it.

For Japan, it also provides a useful forum in which to articulate its interests, and this was, probably, one factor behind the proposal for multilateral security dialogue. This is important because of continuing prospects for instability in the region. One area of potential instability, as mentioned above, is the Korean Peninsula. The situation on the Korean Peninsula has always figured prominently in Japanese security considerations because of its locational proximity. In recent years there has been a partial reconciliation of North and South Korea and Japan has also played a part in reducing the international isolation of the North Korean regime. These developments have, on the one hand, diminished the possibility of overt conflict, but, on the other hand, the suspected North Korean nuclear weapons programme has elevated Western concern about a worsening situation. American Secretary of Defense, Dick Cheney, argued that nuclear proliferation on the Korean Peninsula was the primary threat confronting the United States in the post Cold War era. The fact remains that even though, under North Korean insistence, South Korea is now free of US nuclear weapons,

Pyongyang still refuses to allow inspection of its nuclear facilities as it is obligated to do as a signatory of the Nuclear Non Proliferation Treaty.

We can speculate on how this impasse might be resolved. A worst-case scenario of what might happen is that the American government out of frustration will resort to stronger measures and impose political and economic sanctions. Such sanctions are even now being advocated by members of the US Congress. For example, Congressman Steven Solarz, following a visit to Pyongyang where he discussed the issue of nuclear inspections with the North Korean leader Kim Il Sung, argued that: 'If diplomacy fails, we should then attempt to see if support can be garnered in the Security Council for a resolution mandating comprehensive sanctions against North Korea.'[26]

If a decision to impose sanctions is approved, it will further isolate North Korea from the international community. The Japanese government is unlikely to want this to happen for fear that a desperate North Korean regime might act in a rash and unpredictable manner. The other danger of sanctions is that these may initiate a slippery slide to progressively stronger measures culminating in an outbreak of hostilities, as happened during the Gulf crisis. The outbreak of war in the Gulf came as a real surprise to the Japanese government but a similar sequence of events in Korea would be a greater cause for alarm in Japan. Japan can be expected to urge caution to the United States. The conclusion that follows is that although, in the past, the United States may have been the 'cap in the bottle' of a Japanese military build-up, Japan, within a multilateral regional security structure might in the future begin to exercise a restraining influence on American exuberance and a check on rash decisions.

The significance of this is that it points to changes in the United States–Japan relationship and in the position of Japan within that structure. The United States will have to adjust to a more confident and more assertive Japan even if there is not likely to be a break in that relationship. The adjustment will not be easy. Not only will it require the United States to accept limitations on its freedom of manoeuvre but there may also be problems arising from the fact that there are no clear signposts that might give a clue to the philosophic underpinnings of a more assertive Japanese foreign policy, beyond a general preference for regional stability. Japanese foreign policy has in the past been criticised for lacking an ideology and a coherent set of principles, and none can be discerned as yet. This may frustrate attempts by the US government, and others, to respond to Japanese initiatives and foreign policy behaviour.

IV CONCLUSION

Japan is certain to play a much larger role in the Asia-Pacific region than it has in the past. Japan's primary interests are in ensuring regional stability and it has successfully pushed for the establishment of mechanisms that would allow for broadly based discussions on security. In future, as ASEAN membership increases to incorporate the Indochinese countries and as the number of ASEAN dialogue partners increases, the PMC will become a region-wide and more meaningful forum for promoting security and stability. Within this framework of regional security, the Japanese government is unlikely to want a military role for itself. Instead, its security interests are in promoting both economic interdependence and political reconciliation and reintegration. This explains the various attempts by the Japanese government to break through the cordon of isolation around some of the regional countries. Japan was, for example, the main advocate for lifting the sanctions that had been placed on China in the aftermath of the Tienanmen massacre. The Japanese government is eager to resume full economic interaction with Vietnam despite the American trade embargo that has been maintained pending satisfactory resolution of the issue of American soldiers listed as missing in action in Vietnam. The Japanese government has also maintained channels of communication with the North Korean regime despite its unease about the North's nuclear programme and has even tried to mediate between North Korea and the United States. All these reflect a desire to smooth out troublesome bilateral relationships with a view to creating a stable multilateral environment.

Notes and References

1. N. Tanaka, 'Nihon no kempo taisei "erabi naoshi" no toki' (It is time to 'review' Japan's constitutional structure), *Chuo Koron*, December 1990.

2. 'Nihon gaiko o machi-ukeru shiren' (The trials that await Japanese diplomacy), *Chuo Koron*, February 1992, p. 83.

3. *Mainichi Shinbun*, 18 January 1992. Earlier, in a speech delivered in Singapore on 3 May 1991, Prime Minister Kaifu expressed these same views.

4. Y. Soeya, 'Kambojia wahei to Nihon–ASEAN kankeil' (Cambodian peace and Japan–ASEAN relations), *Kokusai Mondai*, no. 380, November 1991, p. 42.

5. For a general discussion on the notion of global partnership, see Y. Funabashi, 'Japan and America: Global Partners', *Foreign Policy*, no. 86, Spring 1992; R. Holbrooke, 'Japan and the United States: Ending the Unequal Partnership', *Foreign Affairs*, vol. 70, no. 5, Winter 1991–2; and N. Shima, 'Fuhatsu ni owatta Bush ho-nichi no yokkakan' (The failed four-day visit to Japan by Bush), *Chuo Koron*, March 1992. There is also a very different argument which portrays Japan and the United States as rivals and competitors and which sees their bilateral relations more marked by conflict than with partnership.

6. T. Inoguchi, 'Nirion wa ajia to tomo no ajia dakkyaku o hakarel' (Japan and Asia should strive to reach beyond Asia), *Ekonomisto*, 24 December 1991.

7. See *ASEAN Economic Bulletin*, vol. 8, no. 1, July 1991.

8. *Far Eastern Economic Review*, 6 February 1992, p. 50.

9. Chung-in Moon, 'Managing Regional Challenges: Japan, the East Asian NICs and New Patterns of Economic Rivalry', *Pacific Focus*, vol. 6, no. 2, Fall 1991, p. 27.

10. *Nihon Keizai Shinbun*, 28 October 1991.

11. *The Straits Times*, Weekly Overseas Edition, 11 January 1992, p. 2.

12. K. Miyazaki, 'Time to Reevaluate the Security Treaty', *Japan Quarterly*, vol. 37, no. 4, October-December 1990, p. 421.

13. *Canberra Times*, 15 October 1991, p. 1. Commenting on Soviet withdrawal and American retrenchment from the region, a Malaysian defence analyst suggested that China might be tempted to step into the resulting void and that it was necessary, therefore, to maintain a close watch over Chinese actions. See *The Australian*, 25 November 1991, p. 3.

14. *The Australian*, 22 January 1992, p. 9.

15. *The Australian*, 25 July 1991, p. 7.

16. For a discussion of the concept of shared security, see P. Mangold, 'Security: New Ideas, Old Ambiguities', *The World Today*, vol. 47, no. 2, Fall 1991.

17. See *The Monthly Record* (Australian Foreign Affairs and Trade), vol. 62, no. 7, July 1991, p. 403.

18. See *Pacific Research*, vol. 5, no. 1, February 1992, p. 24.

19. *Financial Review*, 25 July 1991, p. 3.

20. K. Fukada, 'Ajia Ka, Amerika Ka: Nichibei anpo naki sekai' (Asia or America: A world without the US–Japan security treaty), *Bungeishunju*, January 1992, pp. 186–7.

21. A. Acharya, 'The Association of Southeast Asian Nations: "Security Com-

munity" or "Defence Community"?', *Pacific Affairs*, vol. 641 no. 2, Summer 1991, p. 161.

22. M. Antolik, *ASEAN and the Diplomacy of Accommodation* (New York: M. E. Sharpe, Inc., 1990) p. 141.
23. *Far Eastern Economic Review*, 1 August 1991, p. 11.
24. *The Straits Times*, Weekly Overseas Edition, 1 February 1992, p. 12
25. *Far Eastern Economic Review*, 6 February 1992, p. 11.
26. *Age*, 24 December 1991, p. 6.

6 Adjusting to the New International Order

Gerald Segal

Japan, like all great powers in the post-Cold War world, is seeking a new role. But for Japan the challenges posed by the new disorder are particularly acute and the internal process of decision-making is especially unsuited to a swift adjustment. Well before the end of the Cold War, Japan knew that it had to become more 'international': that is to say Japan had to become not just an economic superpower, but also take on political and even military roles commensurate with its economic influence. Japan's search for a more normal role, unlike that of Germany which was tied into European alliances, was more lonely, and more dependent on the single bilateral relationship with the United States. In the Japanese context, a more 'normal' role was abnormally hard to define. In the analysis that follows, we identify some of the longer-term geopolitical realities of Japan's international position before turning to a brief analysis of the impact of the major features of change in the new international relations.

I THE GEOPOLITICS OF JAPAN'S POSITION

In our current trendy concern with new world orders, it is often easy to forget some more basic realities that shape a country's options for change. It might be expected that richer and larger states have more choices and control over their own future, but the logic of interdependence often suggests this is not the case. Japan is not only one of the most interdependent nations, but its distinctive geopolitical position seems particularly restricting.[1]

The starting point for a geopolitical analysis of Japan must be that it is a crowded island state. Until 1945, the Japanese had never been successfully invaded, and like in the British case, the result was a curious mixture of exaggerated confidence about their ability to defend their territory and remain aloof from the outside world. Thus the shock of defeat in 1945 and the subsequent American occupation was especially devastating. Not only did Americans administer Japan for a number of years, but Americans even wrote the Japanese Constitution. It was only the Japanese ability to hide real emotions behind an exterior of outward calm that masked the true resentment and shock that was caused by the defeat in the Second World War. In the course of having to live with these dual emotions of defeat and shame, Japanese attitudes to its island status seemed to change. The isolation was no longer simply a cause for confidence about uniqueness and aloofness, it was now a source of insecurity because Japan had come to depend on the United States for its place in the world. The Americans not only provided markets that allowed Japanese industry to revive, but they provided the essentials of Japanese defence policy. Indeed, it could be said that since 1945 Japan ceased to have a defence policy of its own.

The sense of vulnerability as an island was enhanced by the small size of the territory, the fact that its habitable areas were so crowded, and the development of a modern economy in a small space. These features came together after the Second World War when the United States' influence was felt so fully around the country in the process of rebuilding an economy and society. Although Japanese industry was wrecked by the war, the re-building process was not as difficult as in the case of China that rarely had a prewar economic base to remember. The talent and expertise for economic success survived in the minds of the Japanese people and revival was more a matter of opportunity and capital than learning how to live in the modern world.

When, with extensive American assistance, Japan rebuilt its economy, the result was an impressive economy better geared to economic success than might have been the case if Japan had won the war. But Japan also remained aware that it lacked most major natural resources necessary for growth. The consequent dependence on imports required that growth be sustained by massive exports. As Japan grew richer, it also grew more inter-dependent. More than any other major economy, Japan is vulnerable to economic blockade or simply threats to sever indi-vidual lines of contact with the outside world which might threaten what remained a fragile economy.

Of course, by the mid-1980s when Japan became the world's second largest economy, it seemed hard to think of Japan as a 'fragile blossom'. But the geopolitical reality remained in the sense that Japan depended on open access to raw materials from distant parts of the globe and equally open access to equally distant markets in the developed world. What is more, as Japan modernised, its population became even more urban and dependent on the fragile structures of a modern, sophisticated economy. This was not like most states in Asia with large rural and unsophisticated populations. Japanese fragility was also heightened by the fact that the rise to economic superpower status was so swift and tightly managed by the state that the distribution of its fruits were strangely spread. While in terms of per-capita GDP Japan seemed to be richer than Americans or most Europeans, the reality was far less comforting. When calculated in terms of purchasing-power in the late 1980s, the Japanese were merely among the average for OECD states. With half of Japanese homes connected to mains sewers and the average size of homes well below European, let alone American averages, it was clear that too much of the new Japanese wealth was invested in corporate capability and too little in the average consumption of the Japanese people. Economists suggested that as Japan grew richer and greyer in the 1990s, its people would consume more and the economic success would seem less fragile to the average Japanese.[2]

But all this economic growth, even as it became more widely spread, did nothing to decrease dependence on the outside world. Indeed, it did precisely the opposite as Japanese demanded a better quality of life and even the opportunity to travel in greater numbers abroad. The result was so much interdependence that Japan could be said to have had a MAD relationship with the international economy. Mutual Assured Destruction, unlike the military version of the acronym, was less acute, but the reality was unless Japan continued to take part in the international economy, it and the outside world would suffer devastating consequences. Japanese ties took it into the Middle East for energy supplies, across the Pacific and Eurasia for markets, and into East Asia for production bases as the Japanese economy was hollowed out in some industries by rising costs of productivity.[3]

To an important extent, as Japan grew richer, it also diversified its interdependence, at least in terms of export markets. The reliance on the United States declined as markets were found in Europe and Asia. Japanese investment and interest in

Asia was particularly important because Japan had previously been very ambivalent about its relationship to the rest of Asia. Indeed, many Japanese never truly considered themselves to be part of Asia, and the aloofness was enhanced by the rapid economic growth at home while the rest of Asia seemed slow to start. But as a number of East Asians were able to replicate a variation of the Japanese economic miracle, and Japanese funds flowed into these Newly Industrialised Countries, the Japanese attitude began to change. The growing cohort of the new rich in Asia made it easier for the Japanese to see themselves as sharing some problems and opportunities with other Asians.

This is not to say that Japan is really comfortable with being an Asian power. The legacy of its island aloofness is too deeply embedded in Japan and the memories of other Asians of Japanese imperialism is too fresh for there to be real warmth generally in the relationship. But there is real appreciation of the importance of the economic interdependence and, as we shall see below, the retreat of other great powers has left Asia more free to contemplate whether to grow closer to Japan.

Yet for all the obvious good sense in improving Japanese–Asian relations, the Japanese still face important choices of how to orient themselves. In the 1990s the economies of East Asia, Europe and North America will grow into roughly similar sizes. Japanese economic interests will suggest the need to be active in all three parts of the international economy. Eurasia is now shrinking just as the distances across the Pacific seemed to be reduced in earlier decades. Thus the Japan of the early 1990s knew that it had to sustain a basic strategy of interdependence, and probably a complex interdependence with various parts of the globe. In this sense at least, and for all its wealth, Japan remained a fragile country.

But this wealth led many in Japan to believe that they could do something to ease their fragile state if only they could be less abnormally dependent on the economic leg of its power. If Japan could obtain sufficient political and even military influence, surely it would be more able to defend its island and its distant interests. These questions were raised to the top of the political agenda for much of the 1980s, but for a number of reasons they were never resolved. Not only was the Japanese political system especially ill-equipped to take the tough and innovative decisions necessary to alter its international position, but the outside world seemed unwilling to make space for a new Japan. Asians fretted about the consequences of a more militarily active Japan and Europeans and Americans worried about whether Japan was 'mature' enough to cope with greater political responsibility. Under such circumstances, it was easier for Japan to shelter under the American umbrella, keep a low political profile, and get on with an economics-led foreign policy. Even Japanese aid policy was subservient to these principles.[4] But just as the Second World War traumatised Japanese foreign relations, so the ending of the Cold War has forced a reappraisal. While it is too early to tell what choices Japan will make, it is possible to identify what choices it can make.

II   THE NEW INTERNATIONAL ENVIRONMENT

The Disintegration of the Soviet Union

While the disintegration of the Soviet Union in Europe meant that the 'threat' had melted into something like a more diffuse concern with uncertainty and unrest, in East Asia, Russia remained as the obvious successor state with unchanged frontiers. To be sure, a Russia that in political uncertainty, social chaos and economic free-fall did not inspire confidence in a stable future. But at least this sort of Russia showed every prospect of allowing its formidable armed forces in Asia to rust. Japan, which once worried about the Soviet Union's largest fleet in the Pacific and its expanding out-of-area operations, now saw a Russia in retreat and indeed the bear was turning its rump to Asia in order to cope with its major problems in its European part.[5]

The consequent Japanese view of the Russian rump elicited varied reactions in Japan. The more paranoid fretted about the withdrawal of ships from distant operations to home ports opposite Japan's northern island. They noted that Russian hardware had not rusted in one season and the shrinkage of manpower was nothing like as drastic as it was in the European theatres. All of this was true, but it did not necessarily mean that Russia was as much of a threat to Japan as was the Soviet Union. Indeed, there was always serious doubt about whether Japan fully accepted the American notion during the Cold War that there was a major threat from the Soviet Union. Of course, Japan mouthed Cold War slogans and hosted American troops, but it was always important to distinguish between the formal and the real Japanese perceptions.

To an important extent when Japanese used to speak of the 'Soviet threat', they were really talking about what they viewed as Soviet occupation of the Northern Territories – islands taken from Japan in the last hours of the Second World War. The Soviet view was that these were part of the Kurile islands and rightly belonged to them, despite an offer in 1956 to return the most southern ones. In subsequent years, and especially in the late 1980s, the Soviet Union was attracted by the notion of striking a deal if it meant major Japanese financial aid. As the Soviet Union became less Soviet and even showed signs of becoming less of a union, the Japanese of a more conservative persuasion argued for driving a hard bargain and refused any deal unless all the islands were returned. While West Germany got what it wanted from the Soviet Union in 1990 and Eastern Europe was set free the year before, Japan missed its opportunity to strike a deal. The death of the Soviet Union and the European variant of its ideology in 1991 meant that Japan now had to bargain with Russia, and at a time when discussions about a whole range of formerly internal frontiers of the Soviet Union were preoccupying the Russian leadership, the fate of the southern Kuriles were not a priority. Japan had to cool its heels while Russia obtained funds from all G-7 countries, including a Japan that tacitly realised the errors of its previous hardline ways. In the 1990s the stress was far more on offering carrots to Russia in the hope that the disputed Kurile islands could be returned as part of the general reassessment of frontiers. But as time went on, the decentralisation that was even taking place to some extent within Russia meant that local voices in the Russian Far East grew more vocal in the opposition to a sale of territory for aid that would probably sink into the seemingly bottomless pit of the failing economy of European Russia.

Whether Japan receives some of its islands back or not – and it remains likely that a deal of some sort will be done – the new reality is clearly that Japan perceives a much reduced Russian threat. There is certainly no real prospect of war between these two countries, or even of border skirmishes or military tension. There are no large numbers of starving Russians liable to sail to

Japan seeking asylum or food. Nor is there much pressing need to integrate Russia into a regional economy.

In short, Russia has shrunk in political and military size, while its economic potential remains unrealised. Japan, which seeks to increase its political and probably its military role, is aware of the vacuum created by the Russian retreat, but also aware that it might want to seek closer integration of the local Russian economy as a way of improving Japan's own position regarding dependence on distant raw materials. A tame Russia providing the fuel and minerals for the Japanese economy would be better than having to occupy the vast Russian Far East, with all the consequent tensions that would create with China and Korea. Such a tame Russia might eventually also be better than dealing with the more distant resources in Australia and the Middle East.

Russia also looks like continuing the more politically cooperative attitude that took shape under Gorbachev. From a Japanese and even a wider Western point of view, the major gains for regional stability in East Asia were obtained when Gorbachev exerted pressure on the radical regime in North Korea by improving relations with South Korea. Similarly, *détente* with China by 1989 stabilised one of the most dangerous bilateral relationships in East Asia. And of course, Soviet pressure on Vietnam to withdraw from Cambodia was crucial in bringing about a peace settlement in Cambodia and a *détente* between China and Vietnam. In fact, by the time Yeltsin took over the far-eastern reaches of the Russian empire, there was little that the Japanese could have asked him to do on these wider issues that Gorbachev had not already done.

What Yeltsin and his Second Russian Revolution did do was rattle the credibility of the remaining communist regimes in East Asia by destroying the ideological capital in the Kremlin. North Korea and the Indochinese states were especially aware of this damage, although China had long claimed it was on a separate ideological road, albeit in the same neighbourhood. Not that Japan ever found such ideological differences a serious hindrance in developing relations with communist regimes, except in the sense that the ideological label meant that the United States was unwilling to see Japan get too close to states in the rival camp. In any case, there were good reasons why Japan was wary of the bizarre regime in North Korea and the more cynically minded in

Japan were not upset to see a divided Korea. Japanese *détente* with China was 'licensed' by Sino-American *détente* in the early 1970s and the Japanese rapidly became a major entrant through China's Open Door despite the fact that China notionally retained its communist ideology.

In sum, Japan viewed the disintegration of the Soviet union as an unambiguous gain for its own security, even though reservations were expressed about the failure to settle the territorial dispute. But by creating a vacuum of power, Japan was implicitly being asked if it wanted to fill the space. It could do so by sliding into place as a close ally of Russia and help develop its resources in the Far East. Alternatively, Japan could just ignore the open space and hope that no one such as China or Korea filled the gap. In either case, Japan knew that the decrease in Russian power meant that it had far less need for American support in defence against the Soviet Union. Should Japan achieve a territorial settlement with Russia, the perceived need for American assistance will diminish even further. Thus the next most important new feature of the post-Cold War world for Japan is the changed nature of American policy in the Pacific.

The United States in Retreat

As a valued member of the Western alliance, Japan (which was always east of the Soviet Union) had reason to applaud the victory over the Soviet Union and its ideology. But as the Americans began evolving their post-Cold War strategy, Japan looked on with a mixture of awe and worry. Awe, because the United States moved so swiftly in deciding to take troops home, even before the rust on Soviet military equipment had a chance to form. Awe, also because the United States was so swift in putting together a coalition to defeat Iraq in 1991 and revive the United Nations. And Japan was in awe of the way the United States undertook its post-Cold War actions with such little domestic difficulty at a time when the snail-like pace of Japanese decision-making made it hard to undertake even the most initial shifts of policy.[6]

But there was also worry in Japan that the United States lacked caution when it withdrew from Asia as it did from Europe. In Asia, unlike Europe, there were no effective multilateral security or economic structures that might help coordinate policy in the new age. In Asia there were a series of bilateral connections to the United States, but as the United States thinned its security ropes, there was little else to hold Japan in place. At a time when there was a vacuum of power with Russia's retreat, the American action left the Japanese with an even greater sense of drift in their foreign policy.

To be sure, the United States had not withdrawn its nuclear umbrella and its troops had not all gone home. But as the United States continued to cut its forces around the globe and the domestic debate demanded a greater peace dividend, it was not inconceivable that the United States would have no troops based permanently in Asia by the year 2000. Japan had to decide whether it was simply happy for another vacuum to be created around its other flank, and if not, what should be done to fill the space. The United States was also pulling back from its commitments in Southeast Asia and following the previous Soviet withdrawal, there was another vacuum further south which China or even India might seek to fill.

Japan was now facing an opportunity for change, but one that it probably wished had not developed. It was so much easier to have the United States bear defence burdens, especially when Japan never really thought the Soviet threat was very real, especially in southern East Asia. Now the United States was no longer urging Japan to bear more of a defence burden, and in undertaking its withdrawal across the Pacific, the United States was signalling that the choice about taking a more active military role was basically up to Japan. In the short term, the United States was asking Japan to do more, as in the case of the Gulf war. But Washington was also wary about letting Japan fill the new vacuums, especially as other Asians warned about their worries of Japanese militarism. Official American policy was to move to a more flexible and thinly spread force in the Pacific which used smaller facilities around East Asia, and not to withdraw entirely. Americans spoke loosely about the need to fill vacuums, or in the parlance of an older British strategy, to 'hold the ring'. But few could doubt that the American grasp on that ring had loosened and might let slip entirely. There was no escape for Japan – it had to make some tough choices about the kind of order it wanted in East Asia.

Asian Uncertainties

It was not as if the Japanese had no rivals for filling the vacuum in East Asia. Both China and Korea were especially wary about Japan succeeding the United States as the major power in East Asia. China, of course, was the traditional great power of East Asia until the coming of Western imperialism. But while China's very size and wealth allowed it to fend off the Western challenge for far too long, Japan succumbed more quickly and decided to embark on a more comprehensive strategy of learning from the West in order to better compete. When China eventually emerged as a mostly whole country again, it faced Japanese imperialism that could only be defeated with American aid. Even after Japan's defeat, China's communist rulers were only able to produce a military power, while its economy failed to keep pace with that of Japan. When China finally decided to accept reforms of its communist system, Japan was the dominant regional economy and these two Asian powers emerged as more equal rivals than in the old days, albeit with very different kinds of power.[7]

China, like Japan, viewed the end of the Cold War with mixed feelings. Weakening the threat to the north was useful, but it also meant that the United States had less use for China in the anti-Soviet campaign. Nevertheless, China remained a permanent member of the United Nations Security Council, and as a major arms seller and potential proliferator of nuclear weapons, its policies had, if anything, become more important in the post-Cold War world. For these sound geopolitical reasons, China could not be isolated because of its suppression of dissent in 1989, and Chinese approval (or at least avoiding its active disapproval), was critical to the creation of any new world order. Japan, by contrast, was merely expected to help pay for the new order, and not help shape it, at least until it took a more active part in political and military spheres of international relations.

The contrast between China and Japan was an implicit challenge to Japanese lethargy in developing policy for the new world. But China also posed explicit challenges in its pursuit of irredentism. China's determination to take over Hong Kong in 1997, its continuing claim to sovereignty over Taiwan, and its willingness to use force in the South China Sea to take islands it claimed, all posed challenges to a Japan that hoped it would not have to fill a vacuum because no one else would either. China was demonstrating by such action, and also its leading role in the Cambodia settlement, that it would not wait for Japan to extend its influence into regions that China claimed as its own. In essence, the lifting of the superpower overlay merely revealed a Sino-Japanese rivalry in East Asia.

There are a number of distinctive features of this rivalry, most obviously in the sense that Japanese and Chinese power is concentrated in different categories – China in its military punch and Japan in its economy. But it is also distinctive in the sense that although Japanese and Chinese officials both fear the rise of each other, they rarely wish to express such concerns in public, and certainly not in a loud voice. Both couch their worries in terms of a desire to see the United States stay to contain the rival power, but in private conversations decision-makers in both countries will admit that they expect an American withdrawal and therefore a more clear emergence of a Sino-Japanese rivalry.[8]

The manifestations of this rivalry need not be in terms of war. It may well be that as China grows richer it will grow into a more important economic actor in Asia. China may also be in the position to support its irredentist claims and, in so doing, raise Japanese worries that it will have to move soon or face an even tougher China. Should China deploy such offensive equipment as aircraft carriers or in-flight refuelling, Japan is sure to follow suit and break its own prohibitions on acquiring such offensive arms. Should China pursue its claims in the South China Sea which sit astride vital waterways for Japanese resources and trade, or should China threaten Taiwan – an even more important part of the international market economy than China – then Japan is unlikely to sit idly by. At a minimum, an arms race and greater insecurity will develop in East Asia. At worst, there is a risk of more open conflict between China and Japan.

Even if Japan is able to manage its relationship with China, it is sure to have to face a crisis closer to home in Northeast Asia. South Korea, whose economic growth has been more rapid than that of Japan, has had a remarkable string of successes in isolating North Korea and driving it to the negotiating table. Following the end of the Cold War, the reunification of Germany, and Russian and Chinese pressure on North Korea to reform, there is a seemingly unstoppable historical logic that suggests the two Koreas will be reunified in the near future. Whether it comes as a result of the succession to Kim Il Sung in North Korea, or in a crisis surrounding attempts to keep North Korea from acquiring nuclear weapons, it is certain that the process of reunification will be different from what it was in the German case. For one thing, there is no multilateral structure to control the process as there was in Europe, and the North Korean regime is less rational than anything seen in Europe (with the possible exception of the former Albanian regime).

The Korean conditions are also different in the sense that with the exception of Japan, probably all interested parties would probably welcome a reunified Korea. Russia and probably even China would welcome the stability that would bring, while seeing the new unified state as better able to help ward off a rising Japan. Obviously, for both those reasons, Japan might well prefer a divided Korea if the north can be induced to embark on a reform programme that, as in the case of China, is something short of the immediate abandonment of communism. Of course, such a cynical view is not official Japanese policy, but even more so than in the case of French attitudes to Germany, there can be little doubt that there is widespread support for that old French quip, 'I love the Germans so much I am glad there are two of them.'

At a minimum, Japan cannot avoid tough choices about its policy towards Korea. It is also unlikely that it can duck decisions about how to relate to China should it choose to pursue its irredentist claims. Dealing with both of these issues will require new policies towards Russia in the north and Southeast Asia in the south. In short, try as it will to avoid a new vision for East Asia, Japan must face the new realities and the new uncertainties.

The Challenge of Multilateralism

For some, all this discussion of vacuums of power and great power balances smacks of the old international relations. Could it not be that Japan is a new kind of power, better suited to the new forms of international relations? Is it not possible that in fact it is not a matter of international relations as much as just relations on a wider stage? States and their notions of sovereignty are said to be fading in an age of complex interdependence when more than 50 per cent of trade among OECD states takes place within single multinational firms. Japan, with its corporate type of diplomacy may well be the harbinger of the future. It might, but then again there is sufficient evidence that the new international relations has not dawned quite that fast, and in any case Japan is not so well placed to cope even if it did come to pass. The disintegration of the Soviet Union and the new nationalisms in Europe and Central Asia certainly do not suggest that sovereignty is entirely *passé*. But even if it were passing fast in East Asia and other parts of the globe, Japan has not yet provided any real evidence that it can adjust.

If the new notions are that multilateralism on various levels is the new order, then one might expect to find a more active Japanese interest in taking part in such multilateralism. Sadly, the evidence is mixed. It is true that Japan has sought a wider role in the United Nations, including a seat on the Security Council. Japanese officials head the United Nations High Commission for Refugees (UNHCR) and now the peacekeeping operation in Cambodia. Japan is the world's largest aid donor. But Japan has failed to take tough decisions on any important practical matter in these areas. It declined to send even unarmed troops as part of the Security Council-sanctioned operation against Iraq in 1991, and only sent minesweepers after the fighting stopped. Japan failed to take a leading role in the attempts to get a peace settlement in Cambodia, and declined to get out ahead of the cautious American strategy. And worst of all, Japan declined to sanction the dispatch of barely armed troops to United Nations peacekeeping operations in Cambodia. This was not the action of a great power, and certainly not that of an aspiring power anxious to take a more active political and military role. This was not a case of a militaristic Japan considering unilateral action against the wishes of the international community, but rather the abject failure of the Japanese domestic system to produce a consensus for a new policy which would have made Japan a valued member of multilateral collective security. In short, Japan failed the test of multilateralism, although it is bound to try again. Assuming that it can recover its most minimal nerve, then Japan will have to develop the courage to send a decent number of troops and to operations outside of Asia. Not until then will people be convinced that Japan is either willing or able to play a serious part in the new world order.[9]

It may be that Japan has a deeper problem with multilateralism as a form of diplomacy. Could it be that the island mentality is closer to the surface than we thought? Certainly when one looks at the notions of economic cooperation in Asia-Pacific such as Asia Pacific Economic Cooperation (APEC) or the Pacific Economic Cooperation Conference (PECC), Japan seems a reluctant participant, especially when it comes to taking tough decisions instead of merely providing a forum to let off steam. To some extent, Japanese reluctance can be explained by American suspicions about these processes and the Japanese reluctance to upset the United States. To some extent these are simply not important bodies and Japan might feel that especially when there is talk of setting up an East Asian Economic Group, there is no point in antagonising the United States on yet another issue related to trade. But whatever the case, the reality remains an East Asia without any serious multilateralism, with the possible exception of ASEAN, but of course Japan is not a part of Southeast Asia.

If Japan were seeking a careful way to evolve new policies in the post-Cold War world, multilateralism has much to commend itself. But Japan, very much in the shadow of American policy, was unambiguous in its rejection of such ideas in the security sphere. It is true that in 1991 Japan began to develop new approaches to regional arms control and multilateralism, much like the United States. Japanese officials began speaking of a 'multiplex' of security, while rejecting anything that looked like a CSCE for Asia (a CSCA). Some suggested that this multiplex might allow for a more à la carte approach to regional security that focused on such sub-regions as Northeast Asia or even specifically the Korean conflict. Variations on a 2 plus 4 formula for Korea might be useful, but of course the Koreans themselves reject such multilateralism. Perhaps because of such problems, Japan remains among the least interested in developing ideas for regional security and arms control. It has been the Canadians and Australians who have led the way, and perhaps the fact that these are Anglo-Saxon countries tells us something about a wider East Asian cultural aversion to formal multilateralism and arms control. If so, Japan is not so much out of step with its Asian neighbours, even though it is still missing an opportunity to shape a more acceptable world role for a power such as Japan.

III   FINDING NORMALITY IN A DISORDERLY WORLD

There can be no doubt that the time is ripe for major changes in Japanese foreign policy. While some aspects may not change much – such as complex economic interdependence – there is much in the security and political sphere that needs to change. Japan and the world outside can agree that they want to see a more normal Japanese power, but it seems that the Japanese themselves have the biggest problem in thinking about what that might mean in practice. The basic block lies not with the international system, but with Japanese domestic politics. In fact, most of Japan's neighbours would like to see a more normal Japan, but one that is integrated into a more multilateral structure; much like Europeans can cope with a normal Germany, and even an enlarged one, if it is merely part of a series of multilateral arrangements. Should Japan fail to explore such options, it runs the serious risk of taking East Asia back to the future of the 1920s and 1930s.

Notes and References

1. For a general discussion see Takashi Inoguchi, *Japan's International Relations* (London: Pinter, 1991).
2. These issues are discussed in the contrasting Karel van Wolferen, *The Enigma of Japanese Power* (London: Macmillan, 1989), and Bill Emmot, *The Sun Also Sets* (London: Simon & Schuster, 1989).
3. Gerald Segal, *Rethinking the Pacific* (Oxford: Oxford University Press, 1990).
4. Alan Rix, *Japan's Economic Aid* (London: Croom Helm, 1980).
5. For a pre-August 1991 analysis of these issues, see Gerald Segal, *Normalizing Soviet–Japanese Relations*, RIIA Special Papers, London, April 1991.
6. Kent Calder, 'US–Japan Relations: Towards 2000', *The Pacific Review*, no. 2, 1992.
7. Robert Scalapino, 'Japan's Asian Policy', *The Pacific Review*, no. 2, 1992.
8. Allen Whiting, *China Eyes Japan* (Berkeley, Calif.: University of California Press, 1988), and Laura Newby, *Sino–Japanese Relations* (London: Routledge for the RIIA, 1988).
9. Gerald Segal, 'Northeast Asia: New Order or à la Carte Security', *International Affairs*, Autumn 1991.

7 Military Expenditure and Economic Growth: The Case of Japan

Keisuke Matsuyama, Mitsuhiro Kojina and Yutaka Fukuda

I NATIONAL ECONOMY AND MILITARY BUDGET

In the early 1950s the Japanese Self-Defense Forces were established at the request of the United States. This has subsequently been the object of heated discussion as the Japanese Constitution declares that Japan abandons any war potential.

Although Japanese military expenditure was small before 1950, it was not zero. According to Japan's definition of military expenditure, retired servicemen's pensions are not regarded as military expenditure. From the definition of NATO this pension should be regarded as military expenditure. The volume of this expenditure was not negligible. Even today it amounts to 30 per cent of military expenditure in the general budget account. Moreover, before 1950 Japan shared the costs of the Occupation Army, although this was not substantial.

The view that low military expenditure in Japan resulted in high economic growth seems convincing. Instead of sharing economic costs required for the Korean and Vietnam wars, Japan enjoyed only its special procurement boom due to these wars. However, we cannot explain the relationship between economic development and military expenditure well. This relationship is much more sophisticated and complicated.

Enlargement of military expenditure is an obstacle to economic development. If only a small part of military expenditure is forwarded to social welfare, poverty will decline. But the influence of military expenditure on economic development is not always negative. In some cases enlargement of military expenditure does not correspond to a fall in economic development directly.

From the standpoint of cross-sectional analysis, the countries

117

Table 7.1   Time series of comparative per capita measures of GNP and MILEX for selected countries, 1985–8

	Year	GNP per capita	Military expenditure per capita
Korea	1985	2 150	110
	1987	2 690	134
	1988	3 600	147
Singapore	1985	7 420	396
	1987	7 940	463
	1988	9 070	455
Thailand	1985	800	30
	1987	850	32
	1988	1 000	29
Malaysia	1985	2 000	99
	1987	1 810	51
	1988	1 940	99
Indonesia	1985	530	15
	1987	450	8
Philippines	1985	580	8
	1987	590	13
	1988	630	15

Notes:
Unit = Dollar.
Correlation coefficient = 0.9936.

Sources: GNP: *Sekainennkan* (World Yearbook). Kyodotsushinsha military expenditure: *Military Balance* (The International Institute for Strategic Studies, 1988, 1989 and 1990).

in which the ratio of military budget to gross national budget (or ratio of military budget to Gross National Product) is high are not always economically underdeveloped countries. The United States, Britain and France, for example, allocate considerable funds to military expenditure.

This point is exemplified in the time series data and cross-sectional data for GNP per capita and military expenditure per capita for selected Asian/Pacific countries detailed in Tables 7.1

Table 7.2 Cross-sectional analysis of GNP and MILEX per capita
measures for selected countries, 1988

	GNP *per capita*	Military power × 1000 *per capita*
Korea	3 600	14.987
China	330	2.899
Singapore	9 070	20.943
Malaysia	1 940	6.678
Thailand	1 000	4.694
Philippines	630	1.771
Indonesia	440	1.623
Taiwan	7 997	18.182
Brunei	13 282	19.600

Notes:
Unit of GNP = Dollar.
Correlation coefficient = 0.9146.

Military power per capita = $\dfrac{\text{number of military persons}}{\text{population}}$. As the result is very small it is multiplied by 100.

Sources: GNP Population: *Sekainennkan* (World Yearbook), 1990. Military expenditure: *Military Balance*, 1990.

and 7.2, respectively.[1] Neglecting any causality between economic development and military expenditure, we came to the conclusion that at least in Asia increasing national income per capita corresponds to an increasing volume of military budget. In other words, economic growth is necessary to service military power. Clearly, this observation derives from only superficial analysis. But if the Japanese economy is studied from the standpoint of time series analysis a similar phenomenon appears.

Ever since the creation of the Self-Defense Forces the ratio of the military budget to Gross National Budget has been decreasing as a trend. From data collected this ratio to GNP decreased until 1972 while against gross national budget it decreased until 1981. After 1981 the military budget growth rate became higher than the growth rate of any other expenditure. The ratio of military budget to GNP was almost 1.78 per cent in 1955 when the so-called economic growth 'take-off' took place in Japan. Taking account of pension payments for retired ex-soldiers the ratio exceeded 2 per cent in those days. The ratio was highest at the most important time of the Japanese economy, thereafter decreasing as the Japanese economy began to maintain its constant high growth rate. Economic take-off is difficult to achieve. Yet significantly, high military expenditure, by Japanese standards, did not act as an obstacle to the growth process.

II THE MILITARY BUDGET GROWTH RATE

In the previous section we suggested that the military budget's influence on economic progress is not always negative. If we examine this from a different standpoint we come to a different conclusion. But before we examine the relationship between military expenditure and economic progress, we must examine how the volume of the military budget and that of the gross national budget have varied as a trend.

Table 7.3 contains Japanese government data represented by the following symbols:

t ... time
$D(t)$ ... amount of military expenditure at t
$G(t)$ ... the gross national budget

Suppose $D(t)$ and $G(t)$ satisfy the following well-known differential equations:

$$\frac{1}{D(t)} \frac{dD(t)}{dt} = \alpha - \beta D(t)$$

and

$$\frac{1}{G(t)} \frac{dG(t)}{dt} = \alpha' - \beta' G(t)$$

where α, α', and β, β' are proper positive constants. The solutions are

$$D(t) = -M\frac{\alpha}{\beta} \cdot \exp(\alpha t)(1 - M\exp(\alpha t))$$

Table 7.3   Trends of military expenditure and Gross National Product, 1955–91.

Year	Military budget D(t)	Gross national budget G(t)
1955	1 349	9 915
1956	1 429	10 349
1957	1 435	11 375
1958	1 485	13 121
1959	1 560	14 172
1960	1 569	15 697
1961	1 803	19 528
1962	2 085	24 268
1963	2 412	28 500
1964	2 751	32 554
1965	3 014	36 581
1966	3 407	43 143
1967	3 809	49 509
1968	4 221	58 186
1969	4 838	67 396
1970	5 695	79 498
1971	6 709	94 143
1972	8 002	114 677
1973	9 355	142 841
1974	10 930	170 994
1975	13 273	212 888
1976	15 124	242 960
1977	16 903	285 143
1978	19 010	342 850
1979	20 945	386 001
1980	22 302	425 888
1981	24 000	457 881
1982	25 861	496 808
1983	27 542	503 796
1984	29 346	506 272
1985	31 371	524 996
1986	33 436	540 886
1987	35 174	541 010
1988	37 003	566 997
1989	39 198	604 142
1990	41 593	662 368
1991	43 860	703 474

Note: Units $= 10^2$ million Yen.

Source: *Boeihakusyo* (Self-Defense White Paper), (The Defense Agency of Japan, 1970, 1971, . . . 1990, 1991).

and

$$G(t) = -M'\frac{\alpha'}{\beta'}. \exp(\alpha't)(1 - M'\exp(\alpha't))$$

where M and M' are determined from proper initial conditions. It is well known that the parameters of these functions cannot be estimated through applying the least squares method. Instead of differential equations, difference equations:

$$\frac{D(t) - D(t-1)}{D(t-1)} = \alpha - \beta D(t)$$

and

$$\frac{G(t) - G(t-1)}{G(t-1)} = \alpha' - \beta'G(t)$$

are introduced as their approximation. Through applying the least squares method to data, estimated values of α, and α', β and β' are obtained. In order to estimate M, for example, the following equation is useful:

$$M = \frac{D(t)}{D(t) + \alpha/\beta} \exp(-\alpha t)$$

As the estimate of M, we define M as

$$M = \frac{1}{N} \sum_t \frac{D(t)}{D(t) + \alpha/\beta} \exp(-\alpha t)$$

where N is the number of data.

Applying the least squares method to the military data from 1955 to 1991, we have:

$$D(t) = M\ \frac{\alpha}{\beta}\frac{\exp(\alpha t)}{1 - M\exp(\alpha t)}$$

$$\alpha = 0.1254$$

Military Budget
(10^6 million yen)

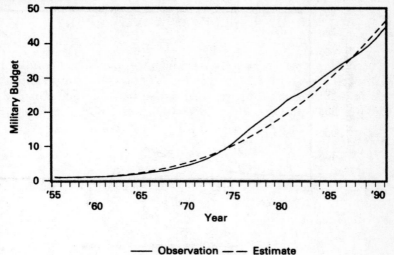

— Observation — — Estimate

Figure 7.1 Logistic model for military budget

β = 1.4701E − 6
M = 0.0112
r = 0.9948.

The related trends are illustrated in Figure 7.1.

Applying the least square root method to the national income data from 1955 to 1991 we have:

$$G(t) = M' \frac{\alpha'}{\beta'} \frac{\exp(\alpha' t)}{1 - M' \exp(\alpha' t)}$$

α' = 0.1764
β' = 1.9291E −7
M' = 6.5677E −3
r = 0.9879.

The related trend is illustrated in Figure 7.2.

In the military expenditure and GNP series, r denotes the correlation coefficient between estimate values and realised (or

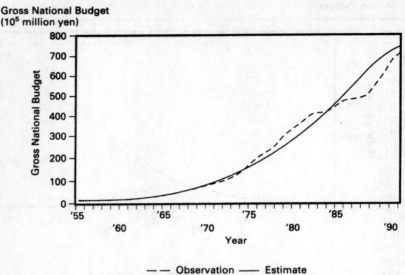

Gross National Budget
(10^5 million yen)

Figure 7.2   Logistic model for gross national budget

observed) values. Also 1955 corresponds to $t = 1$. As $\alpha < \alpha'$ and $-\beta' < -\beta$, we can show that the gross national budget growth rate is greater than that of military expenditure and that factors hindering military expenditure are greater than factors hindering gross national budget.

During the period under examination the ratio of military budget to GNP decreased as a trend. Due to this the strong growth rate of military expenditure does not emerge clearly. Therefore we shall apply the above-introduced method to data from 1970 to 1991 (see Table 7.4). As a result we have:

$$D(t) = M \frac{\alpha}{\beta} \frac{\exp(\alpha t)}{1 + M\exp(\alpha t)}$$

$$\alpha = 0.2030$$
$$\beta = 4.0969\text{E} - 6$$
$$M = 0.8689$$
$$r = 0.9957$$

Table 7.4   Military budget trend, 1970–91

Year	Planned GNP	D(t)	G(t)
1970	724 400	5 695	79 498
1971	843 200	6 709	94 143
1972	905 500	8 002	14 677
1973	1 098 000	9 355	42 841
1974	1 315 000	10 930	170 994
1975	1 585 000	13 273	212 888
1976	1 681 000	15 124	242 960
1977	1 928 500	16 903	285 143
1978	2 106 000	19 010	342 950
1979	2 320 000	20 945	386 001
1980	2 478 000	22 302	425 888
1981	2 648 000	24 000	457 881
1982	2 772 000	25 861	496 808
1983	2 817 000	27 542	503 796
1984	2 960 000	29 346	506 272
1985	3 146 000	31 371	524 996
1986	3 367 000	33 436	540 886
1987	3 504 000	35 174	541 010
1988	3 652 000	37 003	566 997
1989	3 897 000	39 198	604 142
1990	4 172 000	41 593	662 368
1991	4 596 000	43 860	703 474

Note: Units $= 10^2$ million yen.

Source: *Boeihakusyo* (Self-Defense White Papers), 1980, 1981, . . . 1990, 1991.

and

$$G(t) = M' \frac{\alpha'}{\beta'} \frac{\exp{(\alpha' t)}}{1 - M' \exp(\alpha' t)}$$

α $= 0.2596$
β' $= 3.6320E - 7$
M' $= 0.0755$
r $= 0.9882$

In these expressions, 1970 corresponds to $t = 1$. Considering the magnitude of the correlation coefficient the functions estimated are tolerable results of simple time series analysis. As $\alpha' <$

α we can say that the military budget growth rate is less than that of gross national budget. Moreover, as we have $-\beta < -\beta'$ the dampening effect on the military budget is less than on the gross national budget. This is the reason why military expenditure seems to increase with a strong economic growth rate.

In order to estimate the short-term growth rate of military expenditure we shall examine the relation between GNP and military budget. Let $Y(t)$ denote GNP which is estimated from the budget plan authorised by the Japanese Diet. Applying the same method to data from 1970 to 1991 we have:

$$Y(t) = M'' \frac{\alpha''}{\beta''} \frac{\exp(\alpha''t)}{1 + M'' \exp(\alpha''t)}$$

$$\begin{aligned}
\alpha'' &= 0.1741 \\
\beta'' &= 3.183E - 8 \\
M'' &= 0.1080 \\
r &= 0.9901
\end{aligned}$$

and

$$\alpha'' < Min\ (\alpha, \alpha')$$
$$Max\ (-\beta, \beta'') < -\beta''$$

The growth rate of $Y(t)$ is less than that of $D(t)$ and $G(t)$, but the dampening effects on $Y(t)$ are much less than those on $D(t)$ and $G(t)$. In other words, planned GNP increases with a higher growth rate.

A trivial mathematical trick shows:

$$\text{Lim}_{t=\infty} \frac{D(t)}{Y(t)} = 0.01803$$

and

$$\text{Lim}_{t=\infty} \frac{D(t)}{G(t)} = 0.13860$$

Suppose that current trends continue into the future then military expenditure will increase until levels are reached similar to those experienced twenty years ago.

At the same time we have:

$$\text{Lim}_{t=\infty} \frac{D(t)}{2} = 24769.3$$

and

$$\text{Lim}_{t=\infty} \frac{G(t)}{2} = 357415.4$$

In other words, the inflection point of $D(t)$ occurs about 1980 and that of $G(t)$ about 1978. The inflection point for $Y(t)$ does not occur in the period for which we analysed. The growth rate of $D(t)$ began to get smaller in 1980 and so on. From an empirical standpoint, the growth process traces the following steps successively:

Step 1: Preparatory period for growth. The growth rate is almost zero.

Step 2: Take-off period for growth. The growth rate begins increasing.

Step 3: The growth process faces inflection point.

Step 4: The growth rate begins decreasing.

Step 5: Saturation period. No more growth is possible.

The growth rate of $D(t)$ is higher than that of $Y(t)$ and the dampening effects on $G(t)$ are greater than that on $Y(t)$. This is the reason why $Y(t)$ stays in Step 2 and $G(t)$ enters Step 3. However, it should be noted that our analyses have been conducted using variables defined in monetary terms. After 1980, price indices such as the CPI became stable in Japan and thus our analyses of GNP, gross national budget and military budget are not deflated. This should be borne in mind when interpreting the results.

Thus far discussion has been based on the assumption that the status quo holds. But, of course, this may not be the case.

III MILITARY EXPENDITURE AND ECONOMIC GROWTH

Economic growth is mainly affected by the various economic activities of the private sector and the financial and monetary policies of government. The economic conditions of foreign countries also clearly play a part. In order to investigate the impact of military policies on economic growth, consider a quantitative model involving economic growth as a dependent variable and public investment and military expenditure as explanatory variables. Before constructing this model the following crucial problems must be overcome.

The first is that the dependent variable and explanatory variables appear to be correlated with each other through the use of a common deflator. Even though the variables have no relationships the deflator would cause a virtual relationship. The second problem is that both the dependent and explanatory variables have a tendency to increase. This trend results in a high correlation among variables, although any causal relation cannot be presumed theoretically.

To avoid such superficial correlation among variables invoked in the model, variables should be converted to non-nominal ones. Variables which explain stationary processes must be introduced. Accordingly, the following model is developed. The focus of investigation is on the direct and ripple effects of military expenditure on the economy; government fiscal and monetary policies and overseas trade factors are therefore not incorporated. Thus:

$$C(t) = \gamma(Y(t) - G(t)) \tag{7.1}$$

$$G(t) = D(t) + A(t) + \delta Y(t) \tag{7.2}$$

$$\varepsilon \frac{d}{dt}Y(t) = I(t) + \lambda A(t) \tag{7.3}$$

$$I(t) + A(t) = Y(t) - C(t) - \mu D(T) \tag{7.4}$$

where

$$
\begin{aligned}
t &= \text{time} \\
Y(t) &= \text{NNP (Net National Product)} \\
C(t) &= \text{gross national consumption} \\
G(t) &= \text{budget expenditure} \\
D(t) &= \text{military expenditure}
\end{aligned}
$$

$I(t)$ = investment by private sector
$A(t)$ = autonomous investment by government
γ = propensity to consume
ε = accelerator coefficient

The right-hand side of equation (7.1), $Y(t) - G(t)$, refers to disposable income. This is the normal definition of consumption behaviour. The left-hand side of (7.2) can be regarded as government income, while the right-hand side corresponds to the disposal of government income. A hypothesis is introduced which says that the difference between budget expenditure and the sum of military expenditure and autonomous investment is proportional to GDP. Examination of the way budget expenditure is compiled supports the validity of this hypothesis. Equation (7.3) is obtained through modifying the accelerator principle: eliminating the term $\lambda A(t)$, allows $I(t)$ to be regarded as the private sector's induced investment. Supposing that a policy goal is to control undesired economic fluctuation of public investment, then introducing this term into (7.3) may be justified. However, a substantial part of Japanese military expenditure has been appropriated for purchasing and maintaining arms. And as arms are not used as a means of production, they could be classified as consumer durables. Therefore the right-hand side of (7.4) expresses that part of GDP which is left to be consumed. The left-hand side of (7.4) expresses the investments undertaken in both the government and private sectors. Thus (7.4) presumes that saving is equal to investments; that is, GDP is equal to the sum of consumption and investment.

As $Y(t)$, $C(t)$, $A(t)$, $D(t)$ and $I(t)$ are defined in monetary terms, these variables are related to each other through the deflator. It is therefore necessary to introduce stationary variables which are independent of the deflator. Instead of applying an econometric method which estimates each parameter individually, only the relationship GDP and military expenditure is used. Substituting (7.1), (7.2) and (7.4) into (7.3), we obtain:

$$\varepsilon \frac{d}{dt}Y(t) = Y(t) - \gamma\,(Y(t) - D(t) - A(t) - \delta Y(t))$$
$$- \mu D(t) + \lambda A(t) - A(t)$$

that is:

$$\varepsilon \frac{d}{dt} Y(t) = (1 - \gamma + \gamma\delta)\, Y(t) - (\mu - \gamma)\, D(t)$$
$$+ (\gamma + \lambda - 1)\, A(t)$$

This expression is changed into:

$$\frac{1}{Y(t)} \frac{d}{dt} = \frac{1 - \gamma + \gamma\delta}{\varepsilon} - \frac{\mu - \gamma}{\varepsilon} \frac{D(t)}{Y(t)} + \frac{\gamma + \lambda - 1}{\varepsilon} \frac{A(t)}{Y(t)}$$

that is:

$$\frac{1}{Y(t)} \frac{d}{dt} Y(t) = \frac{1 - \gamma - \gamma\delta}{\varepsilon} - \frac{\mu - \gamma}{\varepsilon} \left(\frac{D(t)}{Y(t)}\right)^{*} + \frac{\gamma + \lambda - 1}{\varepsilon} \left(\frac{A(t)}{Y(t)}\right)^{*}$$

$$- \frac{\mu - \lambda}{\varepsilon} \left(\frac{D(t)}{Y(t)} - \left(\frac{D(t)}{Y(t)}\right)^{*}\right)$$

$$+ \frac{\gamma + \lambda - 1}{\varepsilon} \left(\frac{A(t)}{Y(t)} - \left(\frac{A(t)}{Y(t)}\right)^{*}\right)$$

In the above equation, $(A(t)/Y(t))^{*}$ and $(D(t)/Y(t))^{*}$ express trends of $A(t)/Y(t)$ and $D(t)/Y(t)$ respectively. Due to data limitations the above equation is transformed into the following approximation equation:

$$\frac{1}{Y(t)} \frac{d}{dt} Y(t) = \frac{1 - \gamma + \gamma\delta}{\varepsilon} - \frac{\mu - \gamma}{\varepsilon} \left(\frac{D(t)}{G(t)}\right)^{*} + \frac{\gamma + \delta - 1}{\varepsilon} \left(\frac{A(t)}{G(t)}\right)^{*}$$

$$- \frac{\mu - \gamma}{\varepsilon} \left(\frac{D(t)}{G(t)} - \left(\frac{D(t)}{G(t)}\right)^{*}\right)$$

$$+ \frac{\gamma + \lambda - 1}{\varepsilon} \left(\frac{A(t)}{G(t)} - \left(\frac{A(t)}{G(t)}\right)^{*}\right) \qquad (7.5)$$

The following regression equation is introduced so that coefficients of (7.5) are estimated statistically:

$$\frac{Y(t) - Y(t-1)}{Y(t-1)} = P + Qt + M\left(\frac{D(t)}{G(t)} - \left(\frac{D(t)}{G(t)}\right)^*\right)$$
$$+ N\left(\frac{A(t)}{G(t)} - \left(\frac{A(t)}{G(t)}\right)^*\right) \quad (7.6)$$

The left-hand side of (7.6) is measured in real terms, and is published by the EPA (Economic Planning Agency of Japan). Due to the change of statistical bases the real GDP data are available only from 1970. The terms on the right-hand side of (7.6) are defined with either nominal or real statistics. Fractions in (7.6), for example $D(t)/G(t)$, are independent of the deflator.

If $(D(t)/G(t))^*$ is defined employing short-term trends instead of long-term trends and letting $d(t)$ denote the abbreviation of $D(t)/G(t)$, then $d^*(t)$ is defined as:

$$d^*(t) = d(t-1) + (d(t-1) - d(t-2))$$
$$= 2d(t-1) - d(t-2) \quad (7.7)$$

The forecasted trend of d at t is regarded as the value of d at $t-1$ adding the increment of d from $t-2$ to $t-1$.
Therefore:

$$d(t) - d^*(t) = d(t) - 2d(t-1) + d(t-2) = \Delta^2 d(t-2).$$

$(A(t)/G(t))^*$ is defined in the same way. As we cannot find any trend in the left-hand side of (7.6) we assume $Q = 0$. Substituting (7.7) into (7.6) we can apply the least squares method. We obtain:

$$\frac{Y(t) - Y(t-1)}{Y(t-1)} = 4.3913 - 4.6835\ (d(t) - d^*(t)) \quad (7.8)$$
$$+ 0.5681\ (a(t) - a^*(t)$$

$R = 0.6948$, where R is the multiple correlation coefficient. The single correlation coefficient between $(Y(t) - Y(t-1))/Y(t-1)$ and $d(t) - d^*(t)$ is -0.4927. The single correlation coefficient between $(Y(t) - Y(t-1))/Y(t-1)$ and $a(t) - a^*(t)$ is 0.4927. The regression analysis shows that the increase in the

Table 7.5   Growth rate of NNP and military budget, 1971–90 (per cent)

Year	Growth rate of NNP	D(t)/G(t)	A(t)/G(t)
1971	5.2	7.13	17.7
1972	9.0	6.98	18.7
1973	4.7	6.55	19.9
1974	− 0.2	6.39	16.6
1975	4.0	6.23	13.7
1976	4.0	6.22	14.5
1977	4.8	5.93	15.0
1978	5.1	5.53	15.9
1979	5.5	5.43	17.0
1980	3.2	5.24	15.6
1981	3.2	5.13	14.2
1982	3.7	5.21	13.4
1983	2.8	5.47	13.2
1984	4.6	5.80	12.9
1985	4.8	5.98	12.1
1986	2.9	6.18	11.5
1987	4.9	6.50	11.2
1988	5.9	6.53	10.7
1989	4.8	6.49	10.4
1990	5.7	6.23	9.4

Source: NNP: obtained from *Keizaihakusho* (Economic White Paper), (The Economic Planning Agency of Japan, 1980, 1981 . . . 1990, 1991).

military budget has been negatively contributing to economic growth while the increase in public investment has been positively contributing. Thus as expected we can show that increasing military expenditure has a negative influence upon Japan's economic growth. But even so, it is not strong. The correlation coefficient is only about 0.70. In order to examine this relation more closely, we need to analyse the above result from a different perspective.

The ratio of military budget to gross national budget decreases until it amounted to 5.1 per cent in 1981 (see Table 7.5). After that year it began to increase again. Now suppose this ratio had remained at that same level and that the extra amount exceeding this level was switched to public investment. By substituting values which we assume for (7.8), we can simulate the growth rate of NNP after 1981.

The results are:

4.0166	4.6714	4.874	4.4482	4.5163
4.6299	4.1073	4.4197	3.9767	

respectively. Thus, we have:

$$(1 + 0.040166) (1 + 0.046714) \ldots (1 + 0.039767) = 1.471886$$

then NNP (Net National Product) for 1991 is assumed to be 1.471886 times NNP of 1981. Yet actual NNP for 1991 is 1.47974 times actual NNP of 1981.

The results of this simulation show that after 1980 the military budget's strong growth rate had no major influence on the NNP's growth rate. The reason why Japan enjoys its economic growth is not to be found in the policy-making processes of Japan's government. Rather it may be a side-effect of the government's economic or defence policies. The role of the Japanese government in stimulating economic growth is not so strong as foreign observers contend.

IV R&D AND MILITARY EXPENDITURE

In previous sections we showed the increasing tendency for the military budget to have a negative influence on Japan's economic growth. The single correlation coefficient is calculated to be about -0.5. Yet in respect of public investment the Japanese government's budget has a duplicated structure. There firstly is a budget referred to as 'general account'. $A(t)$ was obtained from this (see Table 7.5). There is a further budget called 'Treasury Investment and Loans'. In some cases this budget becomes more than one-third the 'general account'. Public investment by central government is undertaken through $A(t)$ and this budget. Neglecting the effect of 'Treasury Investment and Loans', the single correlation coefficient between the growth rate of NNP and $A(t)$ is almost 0.5. This is very high. On the other hand, the correlation coefficient between the NNP growth rate and $D(t)$ is -0.5. The degree of correlation here is not so strong. In this concluding section this problem is analysed from a different perspective.

Suppose that central government economic policy has no tendency to emphasise military expenditure in the national budget. The ethos of the private (that is, non-governmental) sector will be vitally influenced by this policy. The private sector's economic activities (such as private sector R&D) are oriented to improving the production processes of private enterprise. As a result, the country's industrial structure as a whole improves so that it can produce goods and commodities very efficiently. Economies of scale also become relevant, as rising efficiency leads to increased demand, which in turn facilitates falling unit costs; a virtuous economic cycle being created.

Military industrial products are characterised by high quality. Reliability also needs to be extremely high. And if there is a trade-off relationship between cost and quality, military industries prefer quality to cost. As a result military industrial cost-performance is low. However very high technology is necessary to produce modern weapons. Small-size enterprises cannot afford to develop such technology. Without modern technology it is almost impossible for newcomers to enter the weapons market. Hence, the market for military products has a tendency towards monopoly or oligopoly. If there is no free competition among enterprises manufacturing military products their cost performances become poor. Suppose weapon producing companies are a significant element of the industrial structure then the industrial structure as a whole loses its vitality.

In this chapter we could not completely succeed in showing that low military expenditure represents the sole cause of Japan's high economic growth. The Japanese Self-Defense Forces were created at the request of the United States. To maintain or raise its military power Japan has received various kinds of assistance from America. Gratuitous aid was given until 1969. (Even today, the Japanese Self-Defense Forces receive 'onerous' aid from the United States.)[2] If Japan had not received this aid, then military expenditure would have been greater. However the size of the ratio of military expenditure to GNP or to gross national budget may not be as important for maintaining a high rate of economic growth as we first thought. The attitude of central government, which does not always emphasise the necessity of military power, perhaps ought to be examined in relation to R&D activities.

R&D is the engine, driving technological development so

Table 7.6   Japanese and American shares of R&D and GNP in the totals of five advanced countries (Japan, US, UK, France and W. Germany) (percent)

Year	Japan		USA	
	R&D	*GNP*	*R&D*	*GNP*
1965	4.1	8.4	71.8	63.3
1970	9.7	12.5	66.6	60.6
1975	15.1	16.5	53.1	51.4
1980	17.7	18.6	47.8	46
1983	19.3	19.2	55.7	54.6
1985	20.2	19.4	58.4	57.6

Source: *Trend and Problems of Industrial Technology* (The Ministry of International Trade and Industry (MITI), 1988) p. 62.

crucial to the continued success of an advanced economy. Table 7.6 shows the trends of Corporate R&D and GNP. During the period 1965 to 1985, Japan's share of five-country GNP grew about 2.3 times, while the government's share of R&D grew about 4.2 times. This shows the extent to which Japanese R&D expenditure has grown. In Japan, R&D expenditure has been focused on non-military industry. The Japanese government has been intent on protecting and encouraging domestic high-tech corporations, allowing them to achieve remarkable performances. Japan's R&D level is high in contrast with Western developed countries, and especially so after the first 'Oil Shock'.

It is worthwhile noticing, however, that the share of government R&D is much smaller than for other major countries (see, for example, Table 7.7). The government's contribution to R&D activities has effectively been made to orchestrate competitive conditions rather than financially motivated by short-run profit gain. Although its financial contribution is not great, the government has historically adopted a restrictive approach towards the military industry, while stimulating private sector R&D activities. This results in strong competition among manufacturers of non-military goods. In terms of the nature of modern science and technology, central government plays an increasingly important role both in the field of R&D expenditures and in the field of support and coordination of corporate R&D activities.

By way of a conclusion, while we expect the Japanese govern-

Table 7.7 Governments' share of R&D expenditure (per cent)

Year	Japan	USA	FRG	UK	France
1965	30.8	64.9	47.4	49.3	70.6
1970	25.2	57.0	46.6	51.3	64.3
1975	27.5	n.a.	48.8	52.9	60.1
1980	25.8	47.1	43.7	49.0	56.2
1985	19.4	46.8	39.6	42.6	53.5

Source: *Trend and Problems of Industrial Technology*, MITI, p. 65.

ment to adopt particular 'non-productive' investment, special consideration should be taken to avoid spoiling the market mechanism, the efficient functioning of which is critical to Japan's modern society.

Notes and References

1. GNP and military expenditure were analysed on a per capita basis for the purpose of normalising the data.
2. Japan shares a considerable part of the costs expended by the United States in stationing its Armed Forces in Japan. These monies come from Japan's military budget. In addition, Japan continues to receive what is termed 'onerous' aid from the US Armed Forces. Having regard to utilities and various forms of conveniences, Japan is also obliged to pay for these services.

8 Japanese Defence Industrialisation

Alistair D. Edgar and David G. Haglund[1]

I INTRODUCTION

Outside of the domestic political arena, where it has been a topic of policy debate since the early 1950s, the rebirth and possible expansion of Japan's defence-related, or arms-production, capabilities in the country's industrial base has become an issue of increasing interest only during the past decade.[2] Particularly in the United States, this new interest in the subject of Japanese defence industrialisation results as much from concerns over its potential economic impact on American industry as from analyses of its meaning for the postwar security relationship, or for Japanese defence policy in its own right.

First, the Japanese 'economic miracle' and the relative economic decline of the United States have led both Congress and successive US Administrations, rightly or wrongly, to focus on the massive and persistent annual trade deficits incurred with Japan.[3] In turn, their macroeconomic policy disputes have spilled over into the realm of United States–Japanese defence 'trade', production and technology-transfer agreements. On the US side this has been reflected in the increasing involvement of the Department of Commerce in negotiations over bilateral defence production Memoranda of Understanding (MOUs), traditionally left in the hands of the Defense Department, and in the vocal and effective criticisms of such agreements by Congress. The discussion below of the controversy which surrounded the FSX MOU amply illustrates this spillover.[4]

The second reason for the attention now being paid to Japanese defence industrial capabilities straddles both economic and defence policy issues. Since the early 1980s the Department of Defense (DoD) and Congress have recognised the many problems facing their 'ailing defence industrial base', and the need to consider the implications of a growing military dependence on foreign critical technologies. Even prior to the Gulf War the latter subject was becoming widely debated, and the brief but intense campaign against Iraq highlighted again the importance of Japanese (and other states') technologies for a number of major DoD weapon systems.[5] This dependence could grow further if dual-use technologies or commercial quality control standards and products are adopted in order to reduce procurement costs for the US military in an era of diminishing defence budgets.

Finally, another effect of declining US defence budgets in the wake of the ending of the Cold War has been a careful re-evaluation of America's overseas military commitments, and renewed demands for more equal burden-sharing by its allies. In the case of the United States–Japan security relationship, this has emerged as a call for the latter to do more to monitor and control its sea lines of communication (SLOCs), and to review its potential role in future regional peacekeeping initiatives. Whatever the outcome of these requests the debate over future Japanese defence policy will necessarily raise the question of how to supply, equip and support a possibly more active and independent defence force.[6]

Against the background of this heightened interest our chapter begins by offering an historical review of the Japanese postwar renaissance of defence-related research, development and production capabilities, as well as procurement plans, up to the mid-1980s. Following this review we discuss the current structure, composition and main products of the industry supplier base, and the key industry groups or government agencies that play a role in shaping Japanese defence industrialisation. A subsequent section will touch on Japan's research, development and production strategy (or strategies) in the aerospace industry – including airframe and parts, engines and avionics – through an examination of the key contemporary aircraft programme, the next generation close support fighter, the FSX.

An underlying issue informing our analysis, and one that makes our topic both interesting and difficult, is this: in any conventional understanding of the term, there is no Japanese 'defence industrial base', nor can Japan properly be said to have an 'arms industry'. Japanese companies involved in defence work lack at least three of the principal characteristics generally associated with Western (or 'Soviet') defence industries. Specifically, they lack a significant dependence on:

1. the production of arms or other related products;
2. large government-financed or -supported R&D programmes as the primary incentive for their involvement in defence production; and
3. the export of arms (to increase economies of scale and thus reduce costs).[7]

The final section of our chapter addresses itself briefly to the future prospects for Japanese defence industrialisation. Here, we raise the question of whether the pressures on both the defence-related industries and the Japanese government are likely to lead to the adoption of those activities or policies more closely associated with a 'Western' defence industrial base, or whether the reverse is likelier: namely, that dual-use technologies and commercial standards lead other countries' defence industries to emulate what we might call a 'Japanese model'.

II GROWTH OF DEFENCE PRODUCTION AND PROCUREMENT IN JAPAN

The production of arms or related equipment within Japan had been prohibited by the Allied occupation authorities in September 1945, but by 1947 American weapons already were being shipped to Japan for repair and overhaul as General Douglas MacArthur made use of the skilled, cheap labour force available in that country. The re-emergence of defence production in Japan, however, received its first significant boost when the outbreak of the Korean War in 1950 led the US Military Supply Agency in Japan to begin letting out contracts for the production of ammunition, firearms, vehicles and other supplies to Japanese companies. Between 1952 and 1957 it spent approximately ¥52 billion on these contracts, while the Japan Defense Agency spent an additional ¥1.6 billion on domestic production. By 1957, the Defense Agency had some 38 private companies on its books capable of producing weapons and ammunition.[8]

During this same period the legal basis for the production of military equipment in Japan also was established, with the Law for the Production of Weapons (1 August 1953) and, most important, the United States–Japan Mutual Defense Assistance Agreement (MDAA) of March 1954. The MDAA set out terms for exchanges of 'equipment, materials, services or other assistance' and included reference to American financial and technical support 'of Japan's defence-production industries'.[9] From this agreement Japanese co-production of weapons systems developed in the United States became a key component in the growth of domestic defence industrial capabilities. One scholar, Reinhard Drifte, estimates that over the next 30 years Japan obtained some $10 billion worth of advanced defence-related technology from the United States under the terms of the MDAA.[10]

In July 1954, four months after the MDAA's signing, the Defense Agency Establishment Law and the Self-Defense Forces Law created the Defense Agency and the Ground, Maritime and Air components of Japan's defense forces. The creation of the National Defense Council in 1956 and the adoption of the Basic Policy for National Defense in May 1957 completed the construction of Japan's national security structure. With this structure in place the Defense Agency announced in June 1957 what was to prove to be the first of four plans to build up the military.[11]

The first plan implemented in FY 1958 had a total budget of ¥6.5 billion; its successors in July 1961 and February 1966 grew to ¥13.7 billion and ¥23.5 billion, respectively, although the actual force goals set in these plans were never attained. In each of these first three initiatives the force- and equipment-improvement goals were quantitative in nature, and it was not until the fourth plan in October 1972 that qualitative improvements were set as a key objective, mainly for the Maritime Self-Defense Forces. However, despite a total defence expenditure double that approved for the previous plan, the unexpected international oil crises and rapidly rising inflation severely curtailed the possible impact of this new plan, which again failed to meet its procurement goals.

The difficulties consistently encountered by these defence plans in attempting to fix detailed defence expenditures over several years led the National Defense Council and the cabinet in 1976 to adopt a new approach. The National Defense Program

Outline (NDPO) specified no fixed date for the achievement of its force goals for the Self-Defense Force (SDF).[12] Instead the NDPO, which has continued to form the basic national defence policy of Japan throughout the 1980s up to the present, is complemented with more specific procurement schedules through successive Mid-Term Defense Programme Estimates, better known by the Japanese abbreviation of Chugyo.

The first of these procurement schedules, 56 Chugyo, was established in mid-1979 and ran from 1980 to 1987. Each Chugyo is a five-year plan, though it is reviewed annually and redrawn every three years. Defence equipment scheduled to be procured under the 56 Chugyo was estimated to total approximately ¥4.5 trillion in 1982 prices (about $18 billion), from total defence-related expenditures of some ¥16 trillion (roughly $63 to $64 billion). As with the earlier expansion initiatives, the goals of 53 and 56 Chugyos were not met 'because a Chugyo is not binding on the compilation of the annual defence budget and political conditions made it impossible to make sufficient budgetary allocations'.[13]

Clearly, then, Japanese industrialisation and equipment procurement plans throughout this period from 1950 to 1985 (when the evident shortfalls in 56 Chugyo led the National Defense Council and cabinet to establish new objectives in 59 Chugyo for the latter half of the decade) repeatedly failed to achieve the goals endorsed in any of the various defence plans. Formal and, of equal or perhaps of even greater importance, informal constraints acted to curtail procurement schedules and to circumscribe the associated potential development of a more 'typical' defence industrial base. Before we describe the current structure and chief characteristics of the defence industry, it might be well to consider the nature and scope of these domestic constraints on defence industrialisation in Japan.

Historical Constraints on Japanese Defence Industrialisation

Probably the most widely known formal constraint placed upon the expansion of Japan's military forces – and hence upon its supporting industrial capabilities – has been Article 9 of the 1947 Constitution, the so-called 'no-war' clause.

The continuing strength of the pacifist sentiments Article 9 reflects was at least one major reason why in 1991 Prime Minister Toshiki Kaifu had such difficulty in persuading the Japanese Diet to permit the deployment of peacekeeping forces to the Persian Gulf following Operation Desert Storm. In fact the Diet gave Kaifu permission only to send minesweeping vessels, which were to be used in a limited capacity.[14]

While the no-war clause retains a central place in discussions of the constraints surrounding defence policies and defence industrialisation in Japan, it is important to note that the two paragraphs embodying the clause are the only bona fide constitutional restrictions, and that they do allow for some degree of flexibility. That is, to understand current Japanese defence policies it is essential to analyse Article 9 in light of both domestic public support and the international security climate. Restrictive policies put into place since 1947 – or more precisely, since the 1954 laws establishing the SDF and the Japan Defense Agency (JDA) – are chiefly the consequence of cabinet decisions rather than constitutional legislation.[15]

One related policy practice that was adopted during the same period, and which certainly had a major influence in limiting the postwar rearmament of Japan is perhaps only slightly less well known than Article 9. During the negotiation of the peace and security treaties with the United States, Japanese Prime Minister Shigeru Yoshida argued strongly that the economic revitalisation of his country should be given priority over efforts at rearmament. The 'Yoshida Doctrine' emphasised economic recovery and growth, and reliance on the United States to ensure military security, with the latter often subsumed within a broader concept of 'comprehensive security', as discussed in this book's introductory chapter.[16]

A third important constraint on the further development of Japan's defence industrial base resulted from the involvement of some of the country's industries in supplying the US forces engaged in Vietnam during the 1960s. Potentially massive contracts from the American Department of Defense – reminiscent of the original impetus given by similar orders during the Korean War – and the cooperative attitude of the government raised public concerns over the possibility of Japan becoming embroiled in the Vietnam conflict. Sensitive to such public opposition, in 1967 Prime Minister Eisaku Sato announced restrictions (see Chapter 11 in the present volume) on defence-related exports by Japanese manufacturers additional to the 1949 Export Trade Control Order. Nine years later, in February 1976, strong public protests against the possibility of sales of the C-1 military transport aircraft to governments in the Middle East and Latin America led Prime Minister Takeo Miki to announce additional constraints on defence equipment sales to all other countries, as well as on the export of defence-related technology.[17]

The prohibition on exports – though subject again to interpretation by the government as well as to 'leakage', or circumvention by industry – means that defence production in Japan faces difficult diseconomies of scale. To some degree this problem is resolved by structuring production facilities in such a way as to integrate commercial and defence-related manufacturing, a topic we discuss in the concluding section. However, it has also led some business groups in Japan, most notably the Defense Production Committee of the Federation of Economic Organisations (Keidanren), to issue well-publicised statements favouring a relaxation of export restrictions. Trevor Taylor's chapter in this book examines the subject of arms exports in greater detail, but we also offer a few observations in the concluding section of our chapter, which considers the future prospects of Japanese defence industrialisation.

The last major limitation on defence procurement and production is a non-constitutional policy practice that is nevertheless deeply ingrained in Japanese politics and in public opinion. Subsequent to the decision to scrap the postwar remilitarisation plans and replace them with the National Defense Program Outline, Japanese policy-makers decided that some quantitative limit should be established to restrain defence spending during periods of slower national economic growth. Thus, in keeping with the spirit of the 1947 Constitution and the Yoshida strategy an advisory committee under Michita Sakata – then Director General of the JDA – issued a report in September 1975 recommending that a ceiling of one per cent of GNP be set on future defence budgets.[18]

As Table 8.1 indicates, the 1 per cent ceiling had not in practice been exceeded since the mid-1960s. Following the 1975 report and its endorsement by the cabinet, however, it quickly became a political taboo, which remained untouched until Prime Minister Nakasone began pressing to breach the limit in the mid-1980s.

Table 8.1   Defence expenditures in Japan: selected years, FY 1955 to present

Fiscal year	Defence budget (¥ billion)	General account (%)	GNP (%)
1955	134.9	13.61	1.78
1960	156.9	9.99	1.23
1965	301.4	8.24	1.07
1970	569.5	7.16	0.79
1975	1 327.3	6.23	0.84
1976	1 512.4	6.22	0.90
1977	1 690.6	5.93	0.88
1978	1 901.0	5.54	0.90
1979	2 094.5	5.43	0.90
1980	2 230.2	5.24	0.90
1981	2 400.4	5.13	0.91
1982	2 586.1	5.21	0.93
1983	2 754.2	5.47	0.98
1984	2 934.6	5.80	0.991
1985	3 137.0	5.98	0.997
1986	3 343.5	6.18	0.993
1987	3 517.4	n.a.	1.004
1988	3 700.3	n.a.	1.013
1989	3 919.8	6.50	1.006
1990	4 159.3	6.30	0.997

Source: Japan Defense Agency, *Defense of Japan 1990* (Tokyo, 1990) p. 291.

While Nakasone finally succeeded in breaking the 1 per cent ceiling in 1987 the amount involved was negligible and was achieved only in the face of strong political opposition and public debate. By 1990, expenditures again had fallen below 1 per cent. Despite recent pressures from the United States to accept a greater share of the financial and military burden for national and regional security, the end of the Cold War and the subsequent revision of JDA threat assessments and equipment requirements has reinforced support for the expenditure ceiling and even added a downward impetus. The recent announcement of a 3.87 per cent increase in the JDA draft budget for FY 1992 represents the smallest increase in defence spending for 32 years, and an overall decrease in defence expenditures to 0.941 per cent of GNP, the lowest level since 1981.[19]

III STRUCTURE AND ORGANISATION OF JAPAN'S DEFENCE MARKET

Let us now turn to the current structure and composition of the defence industry supplier base in Japan, and to the roles played by industry organisations and the several governmental agencies that together influence the demand-side factors affecting that base. Associated with most if not all analyses of this subject is an ongoing debate over whether there exists in Japan an influential 'military-industrial complex' pressing for ever greater defence expenditures, and (depending on the nationality and predilections of the particular author) whether a 'growing' arms production capability in Japan foreshadows the emergence of a powerful new competitor in the international defence market.[20] This latter debate we deal with in greater detail elsewhere; here, our comments are directed to the question of structure and composition, for we lack both the inclination and space either to confirm or dispel the notion of a Japanese 'military-industrial complex'. It is our supposition, however, that this concept is really not a useful one for taking the measure of either the structure of, or the relationships within, the Japanese defence market.[21]

As a comparison of Tables 8.2 and 8.3 reveals, the composition of the top 20 Japanese defence contractors has changed considerably between 1950 and the latter part of the 1980s. Only Mitsubishi Heavy Industries, Kawasaki and Japan Steel Works have survived into the current period, with the other companies amalgamating into larger corporate groupings or else abandoning the defence market altogether.[22]

Over the last decade, however, membership of the top 20 defence contractors has stabilised, particularly among the leading 10. As Table 8.3 shows, nine of the top 10 contractors in 1988 had also been in that group in 1983. For FY 1990, the top four contractors by value in Japan were Mitsubishi Heavy Industries Ltd (MHI), which obtained 238 contracts from the JDA worth 440.8 billion yen (US$3.3b); Kawasaki Heavy Industries Ltd (KHI), with 132 contracts valued at 146.4 billion yen (US$1.08b); Mitsubishi Electric Corp., with 268 orders totalling 100.3 billion yen (US$743.3m); and Ishikawajima-Harima Heavy Industries Co. Ltd, which obtained 70 JDA contracts worth some 78.6 billion yen (US$465.2m).[23] It has been argued by analysts who find little evidence to support concerns over a

Table 8.2   The 20 leading Japanese defence contractors (early 1950s)

Rank	Company	Contracts (in US$)
1	Fuji Automobile	37 159 567
2	Japan Steel Works	18 702 862
3	Mitsubishi Heavy Industries	17 113 784
4	Victor Automobiles	8 237 308
5	Shôwa Aircraft	8 215 245
6	Japan Steel	4 375 289
7	New Mitsubishi Heavy Industry	4 048 524
8	Hino Diesel	3 386 675
9	Kawasaki Shipping Industries	2 946 455
10	Japan Steel Construction	1 638 892
11	Japan Shipbuilding	1 424 319
12	Uraga Docks	1 350 279
13	Tokyo-Yokohama Industry	1 297 644
14	Yamamoto Machine Works	883 907
15	Kawa Minami Works	717 593
16	Kyohama Automobile	662 460
17	Kanewa Docks	547 500
18	Yokohama Shipbuilding	491 853
19	Toyota Auto Sales	463 542
20	Sasebo Shipbuilding	432 483

Source: Tetsuya Kataoka and Ramon H. Myers, *Defending An Economic Superpower: Reassessing the US–Japan Security Alliance* (Boulder, Colo.: Westview Press, 1989) p. 64.

'military-industrial complex' in Japan that this large-scale turn-over of defence manufacturers since the 1950s works against the development of powerful lobbying interests. Conversely, it may turn out that the stability in the contractors' ranks over the past decade will lead to a growing voice and increasingly influential role for the country's defence industry.[24]

One enduring characteristic of the Japanese defence market has been that of oligopoly. In FY 1983 the top five contractors obtained 54.1 per cent by value of all orders from the JDA. In FY 1988, as Table 8.3 illustrates, the top five companies received 55.5 per cent of JDA contracts. The top 20 companies together won 73.5 per cent and 75.7 per cent of all JDA contracts in those same years. More recently, for FY 1990, the top four companies by themselves obtained 48.8 per cent of all defence contracts

Table 8.3 The 20 leading Japanese defence contractors FY 1988 (FY 1983 in parentheses)

	Company ranking FY 1988 (FY 1983 in parentheses)	Number of contracts	Amount (¥b)	Total defence contracts (%)
1	(1) Mitsubishi Heavy Industries, Ltd	225	364.2	26.1
2	(3) Kawasaki Heavy Industries, Ltd	130	150.3	10.8
3	(5) Mitsubishi Electric Corp.	236	100.8	7.2
4	(6) Toshiba Corp.	187	83.1	5.9
5	(4) Ishikawajima-Harima Heavy Industries Co. Ltd	66	77.4	5.5
6	(2) NEC Corp.	365	73.6	5.3
7	(7) Japan Steel Works, Ltd	46	31.1	2.2
8	(10) Komatsu, Ltd	69	23.6	1.7
9	(9) Fuji Heavy Industries, Ltd	45	22.1	1.6
10	(14) Fujitsu, Ltd	203	16.8	1.2
11	(20) Oki Electric Industry Co., Ltd	97	16.4	1.2
12	(13) Hitachi, Ltd	71	16.2	1.2
13	(–) Nissan Motor Co.	49	15.1	1.1
14	(15) Daikin Industries, Ltd	66	13.2	0.9
15	(–) Tokyo Keiki Co. Ltd	65	11.4	0.8
16	(–) Shimadzu Corp.	85	10.1	0.7
17	(–) Nihon Kohi, K.K.	80	9.1	0.6
18	(–) Cosmo Oil, Co., Ltd	271	8.2	0.6
19	(–) Kokusai Electric Co., Ltd	169	7.9	0.6
20	(–) Japan Radio Co., Ltd	137	7.9	0.6
	Total	2662 (n.a.)	1058.3 (816.9)	75.7 (73.5)

Sources: Compiled from Kataoka and Myers, *Defending an Economic Superpower*, p. 62; and US Congress, Office of Technology Assessment, *Arming Our Allies: Cooperation and Competition in Defense Technology*, OTA–ISC–449 (Washington, D.C.: US Government Printing Office, May 1990) p. 105.

with MHI alone accounting for almost 30 per cent. Such dominance raises doubts regarding the degree of competition taking place during the process of awarding JDA contracts, and it certainly provides grounds for suspicion on the part of analysts worried about increasing ties and coordination between defence manufacturers and procurement or other policy officials in the government agencies.[25]

This concern may be reinforced by the fact that the major companies, especially HHI and KHI but also Ishikawajima-Harima Heavy Industries, are key participants across several of the product sectors. In shipbuilding, for example, Kawasaki Heavy Industries is involved in the production of steam engines and boilers, diesel engines, and (under licence from Rolls Royce) gas turbines. KHI also produces airframes – though this is dominated by MHI – and aircraft engines for military and commercial aircraft. The largest participant, of course, is Mitsubishi Heavy Industries, which builds 75 per cent of all Japanese-produced military airframes, and is a major competitor in the shipbuilding and the space and missiles sectors.[26]

The importance of these contractors within the defence market, however, must also be seen within the context of their very limited dependence on defence-related business, and the equally limited impact of defence production either within individual market sectors or as a proportion of total Japanese industrial production and employment. While Mitsubishi Heavy Industries for most of the last decade has received a quarter or more of the total value of all JDA contracts, the proportion of MHI's total business accounted for by defence has been approximately 15 per cent. Only six of the top 20 contractors looked to the defence market for more than 10 per cent of business, and for the majority of the others the figure was below 5 per cent. Such figures are a sharp contrast to those for the major defence contractors in the United States and Europe.[27]

Two sectors of Japanese industry that are dependent either heavily or entirely on defence-related contracts are the weapons and ammunition sector (not surprisingly) and the aircraft industry. For reasons that cannot be reviewed here, previous attempts by successive Japanese governments to develop an internationally competitive commercial aircraft industry have proven unsuccessful, whether by domestic research and development or through collaboration with international partners such as Boeing, Pratt & Whitney, and Rolls-Royce. The aircraft industry continues to depend on equipment contracts for the SDF for approximately 80 per cent of total industry sales.[28] This level of dependence, however, is not characteristic of Japanese industry, and the remaining sectors are far less reliant on defence-related production. The shipbuilding industry, which is in a condition of general decline, might consider new naval contracts a boon but still looks to defence for only approximately 5 per cent of total production. The figure is less than one per cent for vehicles, electronics and communications equipment, and other industries (e.g., coal, chemicals, petroleum products, provisions).

A final characteristic of the Japanese defence industry frequently identified by a number of commentators, and used to support a variety of arguments, is the high degree of subcontracting occurring on equipment orders from the JDA.[29] The diffusion of work through subcontracting can be as high as 80 per cent: while the top 20 contractors obtain almost three-quarters of all JDA contracts by value, the Defense Agency has well over 2000 qualified companies registered with its Central Procurement Office. As many as 800 of these firms, or more, can be involved in the production of most defence-related equipment.[30] In the context of the debate over a military-industrial complex, such subcontracting could be argued to demonstrate that a burgeoning galaxy of firms is gaining a stake in supporting defence production and hence in new procurement orders. Alternatively, it also could be used to support the argument that the diffusion of business in such a manner undermines or at least greatly reduces the prospect of oligopoly leading to government–industry collusion.[31]

Defence Procurement and Production: Major Policy Actors

In keeping with the arrangements of the 1947 Constitution and the Yoshida doctrine, the role of the JDA in the process of defence procurement planning – consistent with the position of defence policy itself within the broader scheme of Japanese security policy – is highly circumscribed and tightly regulated.

The Japan Defense Agency (JDA) occupies a secondary station in the overall security bureaucracy It is not a major cabinet position, but rather a secondary state agency. Officials on detail from the Ministries of Foreign Affairs (MOFA), Finance (MOF) and International Trade and Industry (MITI) are involved in key decisions in policy planning and procurement.[32]

In the formulation of its procurement plans the JDA must constantly keep in mind how its policy preferences will enhance (or detract from) its status and bargaining leverage *vis-à-vis* other elements of the bureaucracy.[33] The result of such bureaucratic politics for the formulation and implementation of procurement and production plans – and thus for the defence industry, which fulfils procurement requirements – is discussed in general terms below, and is illustrated more specifically in the case-study that follows this section.[34]

The details of the procurement planning process within the JDA itself have been amply discussed elsewhere and need not be repeated.[35] Of the three internal JDA bureaux that review original procurement requests from the staff offices of the GSDF, ASDF and MSDF the Bureau of Equipment is of most direct relevance to our analysis here, since it is through this bureau that MITI – even at such an early point in the planning process – begins to impress its perspective and interests on the JDA.

The Bureau of Equipment is responsible for input concerning the possible industrial impact of any procurement plans under review, and it must then recommend the *method* of procurement to be undertaken. Should it be off-the-shelf imports, licensed production, co-development or domestic production (or at times a mixture of several of these approaches) MITI's presence is felt in the person of the bureau's Director General who traditionally has served as Director of MITI's Aircraft and Ordnance Division before being seconded to the JDA. This previous post also directly involves MITI in JDA procurement on the basis of the 1952 Law for Enterprises Manufacturing Aircraft and the 1953 Law for Manufacturing Weapons and Munitions. Through these, MITI controls the production of 'all aircraft and parts, weapons and munitions in order to control defence procurement'.[36]

MITI's influence on the formulation as well as the implementation of defence procurement plans and production decisions is likely to continue to increase if the trend towards growing use of dual-use technologies in military systems persists. So far the emphasis on new electronics packages to upgrade existing platforms rather than the production of entirely new platforms or weapons systems in most states' defence budgets, suggests that this trend will not be reversed.[37] The appearance of major Japanese electronics companies such as Toshiba in the top 20 ranks of defence contractors also supports this claim (see Table 8.3).

As was indicated earlier, MITI is not the only bureaucratic actor with its finger in the defence procurement pie. The Ministry of Finance, as is the norm with its counterparts in North America and Europe, casts a critical eye over all JDA equipment or other budget requests, and considers them in the light of overall government budget priorities and outlays. In the context of Japanese defence-related expenditures, however, the MOF's budgetary priorities are set against the history of tight constraints on both the growth and the absolute level of defence spending imposed by the spirit of the 1947 Constitution, the Yoshida doctrine and the continuing public support for at least very close approximation to, if not complete observation of, the 1 per cent GNP ceiling. As one report states: 'while concern with a specific GNP/defence spending ratio is not a major fixation at the ministry, restraining total defence spending is still an article of faith at MOF'.[38]

The remaining bureaucratic actor with interests and influence in defence policies, including procurement, is the Ministry of Foreign Affairs. MOFA is the principal government ministry for establishing security policy in its broadest context, with the general guidelines and priorities of the JDA being set by it. More concretely, however, MOFA can also influence the nature of defence industrialisation through the negotiation of, for example, bilateral technology transfer agreements. The 1983 exchange of notes with the United States on technology transfer (to allow the export of military technologies to the US) was negotiated by MOFA, and has been linked clearly by analysts to the Ministry's broader economic and trade policy interests. The agreement was followed three years later by another, establishing more detailed regulations pertaining to transfers of military technology.[39]

The degree to which the JDA's authority is circumscribed or even superseded in formulating defence procurement plans, and in implementing such plans through specific production policies and contract awards is such that it renders the notion of a compact and influential military-industrial complex problematic. The ability to press one's case with JDA officials may mean less than one's reading of MITI, MOF and MOFA interests. Even then, students familiar with the 'bureaucratic politics' model of decision-making also will recall the uncertainty of outcome from that process.[40] Against the popular misconception of 'Japan Inc.' is the reality that the Japanese government and its several main defence-related bureaucratic actors are by no means a cohesive group, even concerning the comparatively narrow questions of defence procurement and production policies. Before pursuing this last point, however, we should briefly like to examine the role played by the other elements of the so-called military-industrial complex – that is, the industry organisations and major corporations involved in defence production.

As one might expect there are a number of industry organisations representing the interests of particular sectors, the major groups being the Japan Ordnance Association, the Society of Japanese Aerospace Companies, and the Japanese Shipbuilding Industry Association. The members of these groups include older defence-related companies such as HHI and KHI, more recently involved and growing companies such an Toshiba, and producers of purely commercial consumer goods such as Honda and Sony. The principal representative body for defence producers, however, is the Defense Production Committee (DPC), founded in August 1951 and now part of the Keidanren (Federation of Economic Organizations).[41]

The Keidanren is consulted by the JDA at an early stage in the latter's formulation of equipment procurement plans, and thus has access to the decision-making process by which it can inject the views and interests of its members. Drifte draws our attention to the DPC's complaints in 1985 concerning the absence of a coherent Japanese arms industry policy, as well as to the tangled and sometimes contradictory administrative requirements regarding defence production. Several other analysts have also remarked upon industry efforts to persuade the government to relax its prohibition on arms exports.[42]

Thus far, the lobbying efforts of the DPC and other industry organisations appear to have had a not-inconsiderable influence on the details of particular procurement decisions and contracts, but only a very limited effect on the more politically sensitive issues of arms exports or the lifting of the 1 per cent of GNP defence-expenditure ceiling. On these issues the industries themselves are divided, and are especially concerned not to cultivate a poor public image since the bulk of their business is in non-defence areas.[43]

One point on which there does appear to be consensus among governmental organisations and defence industries – although this is hardly either surprising or unique to Japan – is that procurement and production policies should emphasise the use of domestic sources as much as possible. Even before July 1970, when the then Director General of the JDA Yasuhiro Nakasone announced his 'Five Principles' for the development and production of defence equipment, a preference for domestic production was as evident in Japan as it was in the United States or other NATO countries.[44] Discussing the heated controversy surrounding the choice of procurement methods for the next-generation Japanese fighter aircraft, the FSX, one report from the US office of Technology Assessment noted that although there existed different views between MITI, MOF, MOFA and the JDA, and even within the internal bureaus of MITI and the JDA regarding domestic development versus licensed production, the arguments were part of a broader trend:

> There can be no doubt that policy divisions have existed in the past and will continue in the future. However, the long-term trend is toward autonomy at this point and represents a continuum throughout the postwar period.[45]

To complete the main body of our discussion of Japanese defence industrialisation we now turn our attention to a more specific examination of Japanese research, development and production strategies, and present a short case-study of the controversial and widely reported FSX programme.

IV THE NEW FIGHTER AIRCRAFT AND JAPANESE–AMERICAN RELATIONS

As we noted in our introductory section, Japanese defence-industrial base issues have been attracting more attention outside of Japan in recent years. In part this has stemmed from a concern about a potential resurgence of Japanese militarism – a concern we have tended to minimise in our chapter. A more significant reason for the upturn in interest in the Japanese defence-industrial base is to be found in the arena of international trade, especially in the bilateral trading relationship with the

United States. There have been two particularly problematical categories in this regard, each of which can be illustrated by the case-study of this section: the FSX procurement controversy.

The first of these categories relates to the state of the annual deficit run by the United States, currently some $40 billion, in its trade with Japan.[46] Theories abound as to why Japanese–American trade should be so imbalanced, and it is not our purpose here to enter into an explanatory quagmire that has already engendered such a profusion of bitter mutual recriminations, especially since President Bush's Asian visit in early 1992, which one writer has labelled 'The Trip from Hell'.[47] Whatever may be responsible, for instance, for the declining competitiveness of the American automotive sector, there can be no disputing that US defence industry turns out a decidedly 'competitive' range of combat aircraft.

Thus was occasioned in the Washington of the mid-1980s some astonishment and no little bitterness when the Japanese began to intimate that in replacing their soon-to-be obsolete F-1 fighters (the first Japanese-built supersonic warplanes) they might opt for domestic sourcing instead of off-the-shelf procurement from the United States, or at least a co-production/co-development project with the Americans. Anger in Washington was hardly limited to congressional precincts. Indeed, the Pentagon and the Reagan White House, each reasonably disposed both to Japan and to free trade, were by 1987 lobbying the Japanese to drop whatever plans they might have to develop autonomously the FSX (Fighter Support Experimental). It is not hard to understand the Pentagon's position. To begin with there was a concern on 'burden-sharing' grounds about Japan's doing what so many other US allies had done: opting for costly domestic procurement of systems that could be acquired more cheaply from the United States. This was seen to have a potentially injurious impact on US and allied security, because it would further aggravate 'interoperability' problems at the same time as it reduced the real purchasing power of constrained allied defence budgets.

The Pentagon and White House also worried, understandably, about the longer-term potential of Japan developing an arms industry that might at some future date be in a position to compete with American companies in both the domestic Japanese market and in third-country markets, should Tokyo liberalise its policy on arms exporting. A similar concern had triggered the Pentagon's opposition, at roughly this same time, to an Israeli plan to construct a domestic fighter, the Lavi. Since Israel would likely be forced to seek export markets for the Lavi to make it less of a financial burden, American economic interests would be, it was assumed, doubly hurt: US manufacturers would lose sales in Israel itself, and they would find themselves facing competition elsewhere from the Lavi. The Lavi eventually died, in no small measure (but not entirely) a casualty of American opposition to it.[48]

Making the FSX even more of a source of friction than the Lavi were two other contextual factors, specific to the Japanese–American relationship. The first of these, of course, was the bilateral trade imbalance, by 1985 running at about $50 billion a year in Japan's favour. To many Americans in and out of Congress it would have simply defied logic were the Japanese to opt for protected and inefficient domestic production of a combat aircraft at a time when they were making such inroads into the US market for civilian products.

The second problematical category, also related to trade, is the politics of technology transfer in sophisticated weaponry. If the FSX case is a good one for the study of tension between military allies who carry on an unfortunately imbalanced trading relationship, it is no less useful for probing the implications of the growing US dependence upon imported 'dual-use' technology (i.e., know-how at least partly intended or suitable for military applications). Once it was decided that domestic production of the FSX would be inconvenient if not impossible, Tokyo and Washington hammered out a Memorandum of Understanding (MOU) establishing the terms for co-developing what would be an upgraded version of the General Dynamics F-16. The November 1988 MOU called for each country to make available to the other, under a 1983 bilateral sharing accord, certain technologies.

The United States was expected to supply computerised-control technology, as well as some (the amount was disputed) sensitive technology related to the FSX's engines; for their part, the Japanese were to transfer their know-how on composite-wing manufacturing, as well as the technology they were developing on active phased-array radar. Although critics in Japan, especially at MHI, worried about being 'forced' to surrender valuable information about building wings from composite materials, it was really in the United States where opponents of the technology-transfer provisions had the greatest impact. The US critics believed the FSX project would only exacerbate the haemorrhaging of US technical and military skills and secrets, further exposing the country to the twinned dangers of economic underdevelopment and import vulnerability in time of war or national emergency.

Although American foes of the FSX accord (who preferred simply that Japan buy the F-16 or some other advanced craft off-the-shelf) did differ as to their motivation, for our purposes here the most relevant charge they made was that the White House, first under Ronald Reagan, then under George Bush, had continued to demonstrate an agonising inattentiveness to the security implications of trade. Acting without appreciating the extent to which economics and defence were interrelated, the White House had, to the critics, blindly sacrificed US techno-economic interests for the presumed sake of enhanced security and alliance harmony. In effect, charged the critics, the United States would find it had safeguarded neither its security nor its economic interests in permitting – to some, encouraging – the further deterioration of the country's defence industrial base.[49]

Thus for US 'techno-nationalists', the real dilemma symbolised by the FSX arrangement was that it did nothing to halt the erosion of the capabilities on which America's economic and physical security were based. Quite apart from the cost to the economy, the critics stressed, was the potential impact on US defence autonomy in an era in which more and more of the sophisticated components of its high-tech weaponry were being sourced from abroad. Who could know whether in some future crisis Americans might find themselves constrained by the decision of a foreign supplier to say 'no' to US demands for untrammelled access to desperately needed inputs for weapons manufacturing?

V THE 'JAPAN MODEL' OF DEFENCE INDUSTRIALISATION

As our discussion of the growing significance of dual-use technologies indicates, much of the debate over whether there exists

politically powerful military-industrial complex in Japan
misses a fundamental point. It is certainly helpful to begin by
recognising, as does Reinhard Drifte, that Japanese producers
lack a number of key characteristics of a defence industrial base
or 'arms industry' as understood in Europe or North America.
But one must then take the next logical analytical step, and
instead of asking as do many commentators, whether Japan is
developing such a defence industrial base, we should ask if there
might not be apparent an *alternative* path of defence industrialisation, one that we might label the 'Japan model'? Support for
such a notion can be taken from similar arguments to be found in
the broader literature on international relations, where Japanese
postwar economic growth is said to demonstrate the 'rise of the
trading state' or the development of the 'new civilian powers'.[50]

The principal characteristics of this Japan model are the
emphasis on the development and application of dual-use technologies such as semiconductors, microelectronics and robotics
for defence as well as commercial products; the dominance of
private-sector commercial R&D, with the cliche of 'spin-offs'
from defence R&D into civilian sectors being replaced by 'spin-
on' in the opposite direction; the increasing acceptance of commercial product specifications or quality standards for use in the
defence sector; and the widespread integration of defence and
civilian production within single manufacturing plants, even off
the same equipment and production lines.[51]

If we look to the international armaments market and to the
defence manufacturers of the United States and Europe, the last
half-decade in particular has shown trends towards the adoption
of these characteristics, or at least increasingly serious attention
to their possible utility. The growing importance of dual-use
technologies has led to numerous government, industry, and
military studies and some well-known legislative initiatives –
studies and initiatives that frequently refer to the new phenomenon of 'spin-on'. Moreover, the US Department of Defense
(DoD) has been pressed to expand its use of commercially
developed and tested products wherever feasible in order to
reduce costs. Many American corporations involved in both
civilian and defence production are at the same time seeking to
reduce their defence exposure and are voicing complaints over
the inefficiencies resulting from their need to observe strictly
enforced DoD regulations maintaining the complete separation

of defence R&D, production facilities and even workforces from those of their commercial operations.[52] These trends are being further reinforced as declining defence budgets throughout NATO, and a smaller and more competitive international market affect orders for new weapons platforms and major systems.[53]

We do not suggest that the 'Japan model' of defence industrialisation is entirely new or unfamiliar. The policy of *kokusanka*, or preference for domestic development and production over other forms of international collaboration and procurement, has well-documented and firmly entrenched counterparts in the policies of all of the major NATO states, as well as many of the smaller ones.[54] However, we do suggest that it is necessary to consider carefully how we look at questions relating to defence production in Japan.

The example of the export issue may be used to illustrate our point here. It is by no means clear whether advocates of increased Japanese defence exports (or critics in the West) will ever overcome the very strong barriers that they face, both formal and informal. However, if the increasing emphasis on dual-use technologies does offer growing export opportunities to Japanese companies and they are able to take advantage of this market, it may be more accurate to discuss this not in terms of the adoption in Japan of a 'Western' defence industrial base but rather as a result of the changing nature of technology and of the international defence market. In that case perhaps debates over military-industrial complexes and exports should be superseded by consideration of the lessons of Japanese defence industrialisation for Western manufacturers seeking new corporate strategies in the 1990s.

Thus it really does not so much matter that Japan is not becoming like the Western defence industrial powers; much less is it significant that Japan will not breach the 1 per cent-of-GNP ceiling. Increasingly, the technology of weaponry, and the processes through which states procure their armaments will make Japan of necessity, and perhaps even willy-nilly, a more dominant factor in global defence production. It simply cannot avoid this, even if it would wish to.

Notes and References

1. The authors would like to acknowledge the research support they have received from the Centre for Studies in Defence Resources Management, National Defence College, Kingston.

2. For a report that summarises the reasons for this growing interest within the United States, and also provides useful references to previous studies, see US congress, Office of Technology Assessment, *Arming Our Allies: Cooperation and Competition in Defence Technology*, OTA-ISC-449 (Washington, D.C.: US Government Printing Office, May 1990). A European analysis of Japan's potential to develop as a competitive arms exporter, is Jean-Marc Domange, *Le Réarmement du Japon: Le Japon peut-il devenir un exportateur d'armement?* (Paris: Fondation pour les Etudes de Defense Nationale, 1985).

3. General recent commentaries on US–Japan economic relations include Michael Mastanduno, 'Do Relative Gains Matter? America's Response to Japanese Industrial Policy', *International Security*, vol. 16, no. 1, Summer 1991, pp. 73–113; and Commission on US–Japan Relations, *The 1990s: Decade of Reckoning or a Decade of a New Partnership?* (Washington, D.C.: Department of Commerce National Technical Information service, March 1991).

4. The combination of national economic, political and technological interests that surfaced in the FSX project is discussed in David G. Haglund, with Marc Busch, '"Techno-Nationalism", and the Contemporary Debate over the American Defence Industrial Base', in *The Defence Industrial Base and the West*, David G. Haglund, ed. (London: Routledge, 1989) pp. 234–77.

5. A comprehensive, though now dated expression of anxieties over the condition of the US defence industry is US Congress, House Committee on Armed Services, *The Ailing Defense Industrial Base: Unready for Crisis*, 96th Congress, 2nd session (Washington, D.C.: US Government Printing office, 1980). A more recent review is Report to the Secretary of Defense by the Under Secretary of Defense (Acquisition), *Bolstering Defense Industrial Competitiveness: Preserving Our Heritage, Securing Our Future* (Washington, D.C.: Department of Defense, July 1988). An interesting examination of how to specify and operationalise the concept of foreign dependence is provided by Theodore H. Moran, 'The Globalization of America's Defense Industries: Managing the Threat of Foreign Dependence', *International Security*, vol. 15, no. 1, Summer 1990, pp. 57–99. More specific reviews of the issue may be found in Report to the Chairman, Subcommittee on Technology and National Security, Joint Economic Committee, US Congress, *Industrial Base: Significance of DoD's Foreign Dependence*, GAO/NSIAD-91-93 (Washington, D.C.: United States General Accounting Office, January 1991); and Report to the Honourable Helen Delich Bentley, House of Representatives, *Defense Procurement: DoD Purchases of Foreign-Made Machine Tools*, GAO/NSIAD-91-70 (Washington, D.C.: United States General Accounting Office, February 1991).

6. A useful discussion of the efforts of Prime Minister Nakasone to engage Japan in a more active defence policy and in regional security issues is

provided by S. Javed Maswood, *Japanese Defence: The Search For Political Power* (Singapore: Institute of Southeast Asian Studies, 1990). Such efforts inevitably also raise concerns amongst other Asian states, as highlighted in George Leopold and Naoaki Usui, 'Japan Policy Threats Revive Old Fears', *Defense News*, 30 September 1991, pp. 17 and 22.

7. These characteristics are suggested in Reinhard Drifte, *Arms Production in Japan: In Military Applications of Civilian Technology* (Boulder, CO: Westview Press, 1986) p. 3. The concept of a defence industrial base is discussed in Haglund, *Defence Industrial Base*, pp. 1–2. A very valuable contribution to analyses of defence industries from a US perspective, of course, is Jacques S. Gansler, *The Defense Industry* (Cambridge, MA: MIT Press, 1980).

8. From Drifte, *Arms Production in Japan*, pp. 9–10, and Totsoya Kataoka and Ramon H. Myers, *Defending an Economic Superpower: Reassessing the US–Japan Security Alliance* (Boulder, CO: Westview Press, 1989) pp. 55–6.

9. 'Mutual Defence Assistance Agreement Between Japan and the United States of America (Excerpts)', Japan Defense Agency, *Defense of Japan 1990* (Tokyo: Defence Agency, 1990) p. 300.

10. Drifte, *Arms Production in Japan*, p. 11.

11. The figures below are taken from Drifte, *Arms Production in Japan*, pp. 13–17; and Kataoka and Myers, *Defending an Economic Superpower*, p. 56.

12. The force goals set out in 1976 remain those of the current Self-Defence Force. See, for example, Japan Defense Agency, *Defense of Japan 1990*, p. 252.

13. Drifte, *Arms Production in Japan*, p. 16.

14. On the 1947 Constitution and its continuing relevance, see Maswood, *Japanese Defence*, pp. 2–3; the entire text considers the continuing strength of Japanese pacifist sentiments into the 1980s. The current Prime Minister, Kiichi Miyazawa, has been defeated in his recent attempt to pass legislation permitting overseas deployment of troops for peacekeeping operations: see Naoako Uiui, 'Miyazawa Loses a Round', *Defense News*, 16 December 1991, p. 3.

15. Office of Technology Assessment. *Arming Our Allies*, p. 108.

16. Maswood, *Japanese Defence*, pp. 38–42. See also Kenneth B. Pyle, 'Japan, the World, and the Twenty-first Century', in *The Political Economy of Japan*, vol. 2, *The Changing International Context*, Takashi Inoguchi and Daniel I. Okimoto, eds (Stanford: Stanford University Press, 1988) pp. 454–7.

17. Drifte, *Arms Production in Japan*, pp. 73–4; and K. V. Kesavan, 'Japanese Defence Policy Since 1976: Latest Trends', *Canberra Papers on Strategy and Defence*, No. 31 (Canberra: The Australian National University, 1984) pp. 24–5.

18. Maswood, *Japanese Defence*, pp. 28–33, offers a thoughtful background on the development of postwar Japanese defence budgets and the 1 per cent GNP ceiling.

19. See Kensuke Ebata, 'Japan's Spending Curbed for FY 92', *Jane's Defence Weekly*, 11 January 1992, p. 41.

20. Support for the thesis that a growing defence industry lobby exists in Japan capable of exerting pressure on legislative restrictions against arms exports, can be found in Drifte, *Arms Production in Japan*, especially chapters 6 and 7. The opposite view is expressed by Kataoka and Myers, *Defending an*

Economic Superpower, chapter 5. Also see Domange, *Le Réarmement du Japon*, especially chapter 10.

21. We have been interested to note that existing analyses of the structures of defence-related industries, of government–industry relations, and other issues in Japan generally have not considered whether the terminology they employ may in fact be appropriate in this context at all. Although our alternative description is by no means complete, we consider it at least of some value to question whether the process of defence industrialisation in Japan could be taking an entirely novel direction, a 'Japan model'. This theme is developed further in the conclusion of our chapter.

22. Kataoka and Myers, *Defending an Economic Superpower*, pp. 61–5. MHI, KHI, and Japan Steel Works, of course, are key manufacturers with historical links to pre-1940 Japanese industry: see office of Technology Assessment, *Arming Our Allies*, p. 104.

23. Naoaki Usui, 'Mitsubishi, Kawasaki Lead Japan in Defense', *Defense News*, 22 July 1991, p. 21.

24. A complicating factor here, however, is that it is difficult to decide the extent to which greater flexibility in the interpretation of, for instance, arms export and technology transfer restrictions by the Japanese government is the result of effective defence industry lobbying, versus the outcome of strong pressures from the United States government. On the latter, see Kesavan, 'Japanese Defence Policy Since 1976: Latest Trends', *Canberra Papers on Sociology and Defense*, no. 31, Canberra: The National University, 1984, pp. 22–32.

25. Office of Technology Assessment, *Arming Our Allies*, p. 104.

26. Drifte, *Arms Production in Japan*, chapters 3 to 5, provides an examination of the shipbuilding, aircraft, and the space and missile sectors of the Japanese defence industry. See also, for example, Daniel Sneider, 'Mitsubishi Heavy Industries Tops List of Japanese Defense Firms', *Defense News*, 15 May 1989, p. 44.

27. An instructive comparison is provided in Kataoka and Myers, *Defending an Economic Superpower*, pp. 62–3. While KHI in 1983 looked to defence for 16.9 per cent of total sales, and MHI for 10.8 per cent, the figures for General Dynamics Corporation, McDonnell Douglas Corporation, and Grumman Corporation were 80 per cent, 70–80 per cent, and 90 per cent, respectively. Over-exposure to defence remains a difficult problem for US companies in the early 1990s.

28. See Drifte, *Arms Production in Japan*, p. 24; Kataoka and Myers, *Defending an Economic Superpower*, p. 57.

29. Office of Technology Assessment, *Arming Our Allies*, pp. 64–6, argues that the high degree of subcontracting provides Japanese companies with an opportunity to diffuse new technologies acquired from licensed production agreements rapidly throughout both defence and commercial industries.

30. Kataoka and Myers, *Defending an Economic Superpower*, pp. 61–4; and Drifte, *Arms Production in Japan*, pp. 23–5, and footnote 18, p. 29.

31. These arguments may be found in the two texts cited above, and in Kesavan, *Japanese Defence Policy Since 1976*, p. 31.

32. Kataoka and Myers, *Defending an Economic Superpower*, p. 65.

33. Although the OTA analysis suggests that there is a vigorous industry lobby

with close ties to the Japanese government, it also notes somewhat para-doxically that there is a strong, competitive rivalry between sections of the bureaucracy, and even within individual branches, over the budget process and procurement plans. Office of Technology Assessment, *Arming Our Allies*, pp. 102–3.

34. That defence procurement is an intensely political process will come as a surprise to nobody familiar with the wranglings frequently associated with US defence budgets or those of the other NATO governments. For a text which examines this topic across several states, see Haglund, *Defence Industrial Base*.

35. Kataoka and Myers, *Defending an Economic Superpower*, pp. 65–7.

36. Ibid, pp. 67–8.

37. Drifte, *Arms Production in Japan*, pp. 35–41.

38. Office of Technology Assessment, *Arming Our Allies*, p. 103.

39. Ibid, p. 102. The text of the note is reproduced in Japan Defense Agency, *Defense of Japan 1990*, p. 301.

40. Further reviews of the making of defence and foreign policies in Japan which illustrate the complex nature of the process are Shuzo Kimura, 'The Role of the Diet in Foreign Policy and Defense', in *The Japanese Diet and the US Congress*, Francis R. Valeo and Charles E. Morrison, eds (Boulder, CO: Westview Press, 1983) pp. 99–111; and Donald C. Hellmann, 'Japanese Politics and Foreign Policy: Elitist Democracy Within an American Green-house', in *Political Economy of Japan*, vol. 2, pp. 345–78.

41. Drifte, *Arms Production in Japan*, pp. 25–6.

42. Ibid, p. 26. Also, Office of Technology Assessment, *Arming Our Allies*, pp. 70 and 106; and Caleb Baker, 'Japanese Defense Firms Expect Robust Decade', *Defense News*, 18 February 1991, p. 16.

43. Office of Technology Assessment, *Arming Our Allies*, p. 69.

44. The OTA report discusses differences of opinion within the Japanese bureaucracy over the emphasis on domestic production versus inter-national collaboration, ibid, pp. 66–8. Nakasone's statement of Japanese R&D and production policies is discussed in Drifte, *Arms Production in Japan*, pp. 12–13. More specific statistical details of domestic and foreign equipment purchases can be found in Japan Defence Agency, *Defence of Japan 1990*, p. 318.

45. Office of Technology Assessment, *Arming Our Allies*, p. 68.

46. David E. Rosenbaum and Keith Bradster, 'Candidates Playing to Mood of Protectionism', *New York Times*, 26 January 1992, pp. 1, 2.

47. Jim Hoagland, 'For Bush, One More Stumble', *International Herald Tribune*, 10 January 1992, p. 6.

48. See, for a comprehensive discussion, Galen Roger Perras, in Haglund, *Defence Industrial Base*, pp. 189–233.

49. Stephen Kreider Yoder, 'Japan, US Agree on Jet Based on F-16, But Technology Transfer Debate Persists', *Wall Street Journal*, 30 November 1988, p. 16.

50. The latter term is used in Hanns W. Maull, 'Germany and Japan: The New Civilian Powers', *Foreign Affairs*, vol. 69, no. 5, Winter 1990/91, pp. 91–106. The fullest discussion of the notion of trading states is found in

Richard Rosecrance, *The Rise of the Trading State: Commerce and Conquest in Modern World* (New York: Basic Books, 1986).

51. See Drifte, *Arms Production in Japan*, pp. 35–41, 58–62 and 79–83; and Office of Technology Assessment, *Arming Our Allies*, pp. 63–6.

52. Ibid, Arming Our Allies pp. 24–8 and 77–85. For useful reviews of corporate diversification and other strategies, see Laurens van den Muyzenberg and Godfrey Spickernell, *Restructuring in the Defence Industry*, A Report by HB Maynard Management Consultants, undated; and Bill Pietrucha, 'Contractors Cautiously Diversify Product Lines', *Defense News*, 22 July 1991, p. 20.

53. The trend towards reduced numbers of new orders was recognised prior to the effects of more recent and domestic cuts in national defence budgets. See Thomas A. Callaghan, Jr., *Pooling Allied and American Resources to Produce a Credible, Collective Conventional Deterrent* (Washington, D.C.: Department of Defense, 1988).

54. A recent discussion of this preference is Ethan Barnaby Kapstein, 'International Collaboration in Armaments Production: A Second-Best Solution', *Political Science Quarterly*, vol. 106, no. 4, Winter 1991–2, pp. 657–75.

9 International Arms Collaboration: Euro-Japanese Collaboration in Aerospace

Keith Hartley and Stephen Martin[1]

I INTRODUCTION: THE POLICY ISSUES

Three economic powers, namely the EC, Japan and the USA dominate the world economy. With a GDP of about US$500 billion each in 1989 the EC and the USA are the world's two largest markets. Although Japan's GDP is substantially less than this, Table 9.1 reveals that it is still equal to the GDP of the three largest EC member states combined. Moreover, income per head in Japan is greater than anywhere in the EC or the United States. Japan's military expenditure is constrained to 1 per cent of GDP. Because Japan's GDP and the proportion of it allocated to defence are relatively small, Japanese defence expenditure is less than that in both the EC and the United States. However, as Table 9.1 reveals, in absolute terms Japanese defence expenditure is similar to that of any of the individual EC member states.

The market for defence equipment is dominated by high technology, particularly in the aerospace and electronics sectors. Independence in defence procurement is costly and, in real terms, the cost trends for successive generations of combat aircraft are upwards. For example, the total development costs of the four-nation European Fighter Aircraft are estimated at over £7.5 billion and unit production costs at some £30 million per copy (1991 prices).[2] And since 1945, the real unit production cost of UK combat aircraft has increased at an average rate of about 8 per cent per annum, so resulting in a nation buying smaller numbers of each type produced.[3] On this basis, supporting a national defence industrial base is expensive and there are alternative and sometimes lower-cost methods of purchasing equipment.

164

Table 9.1 Comparative economic and military indicators for Japan, EC and the USA, 1989

Country	GDP (US $bn)	Population (mn)	Per capita (GDP US $)	Military expenditure (US $mn)	Armed forces (000s)	Weapon procurement (US $mn)
Japan	2 967	123.1	24 097	29 491	249	8 219
FRG	1 240	61.6	20 112	35 008	503	6 652
France	990	56.2	17 635	36 494	554	9 746
Italy	859	57.5	14 932	20 559	533	4 091
UK	837	57.2	14 628	34 292	318	7 830
Spain	362	39.1	9 248	7 583	277	1 065
Netherlands	237	14.8	16 001	6 791	106	1 196
Belgium	161	9.9	16 336	4 035	110	400
Denmark	108	5.1	21 131	2 263	31	296
Greece	46	10.0	4 596	3 116	201	682
Portugal	36	10.5	3 457	1 415	104	168
Ireland	31	3.5	8 917	488	13	41
Luxembourg	8	0.4	21 579	79	1	3
EC 12	4 916	325.8	15 220	152 123	2 750	32 170
USA	4 959	249.4	19 884	289 149	2 241	73 244

Note: Figures are in constant (1988) prices.

Source: *SIPRI Yearbook 1991 World Armamants and Disarmament* (Stockholm International Peace Research Institute, Oxford, 1991) p. 136.

Instead of independence, a nation might prefer to import equipment. Or it might prefer to manufacture foreign aircraft under licence or on a co-production basis, or to undertake an offset deal or to enter into a collaborative programme involving joint development and production.[4] Japan's involvement in collaborative defence projects with nations other than the United States is subject to policy constraints: Japanese firms are only allowed to transfer military technology to the United States. However, the increasing development of dual-use technologies might mean that, in future, this will become less of a barrier. Laws and constitutions can be changed and in the long run anything is possible. On this basis, then, Euro-Japanese arms collaboration is a feasible long-run policy option for Japan. There are economic and military links between Europe and the United States, on the one hand, and between Japan and the United States on the other; but less so between Europe and Japan. Euro-Japanese arms collaboration would increase military and economic ties between the two groups whilst reducing

their dependence on the United States for high technology products. Such ties might also reduce the probability of future military conflicts.[5]

This chapter analyses the Japanese and European aerospace
industries; it reviews collaborative programmes and concludes
with an assessment of the prospects for Euro-Japanese collaboration. It is recognised that the chapter is speculative, outlining a
possible future scenario.

II   THE JAPANESE AEROSPACE INDUSTRY

Like Germany, Japan was prohibited from building aircraft until
1955 and Japanese policy still prohibits weapons exports to any
country except for the transfer of military technology to the
United States. As with Germany, the Japanese aerospace industry has grown through licence agreements, largely with US manufacturers. But unlike Germany, Japan has not supplemented
its licensed production with participation in collaborative ventures. Another difference between the two countries is Japan's
enthusiasm for the licensed production of American-designed
helicopters whereas West Germany has concentrated on fixed-
wing aircraft.[6]

Table 9.2 shows the growth and relative size of Japan's aerospace industry over the period 1979–89 in terms of the numbers
employed in the industry. Compared with the United States, the
UK, France and the FRG, Japan's aerospace sector is relatively
small. However, as Table 9.3 reveals, the recent growth in
aerospace turnover has been impressive, at almost twice the rate
of the UK and some 40 per cent higher than that of the EC as a
whole. Moreover, as Table 9.2 shows, labour productivity in the
Japanese aerospace industry exceeds levels achieved by the EC
and North America.

Table 9.4 reports the division of aerospace activities by subsector and market destination for the three major economic
blocs. Compared with the EC and the United States, the
Japanese aerospace industry derived a relatively high proportion
of its turnover from space activities in 1989. This reflects the
importance attached to the space programme and, in particular,
the development of an independent satellite launch capability.

Table 9.2   Employment in the aerospace sector in Japan, EC12 and USA

Country	1979	1984	1989	Growth 1979–89 (%)	Labour productivity (000s ECUs per employee) (1989)
Japan	25 653	25 986	28 639	11.6	188.7
Belgium	6 272	5 757	5 239	− 16.5	59.4
France	106 297	127 815	120 334	13.2	98.6
FRG	60 866	65 366	94 456	55.2	89.2
Italy	38 370	42 885	50 500	31.6	58.9
Netherlands	7 935	9 055	12 924	62.9	83.5
Spain	7 335	11 157	12 581	71.6	47.7
UK	196 566	203 083	193 911	− 1.4	65.6
EC 12	423 637	465 118	489 945	15.7	77.5
Canada	37 700	42 300	63 632	68.8	88.2
USA	842 000	817 000	992 000	17.8	118.2

Note: Labour productivity is calculated as turnover per employee.

Source: Commission of the European Communities, *The European Aerospace Industry: Trading Position and Figures 1992*, EC – DG111 (Brussels, 1992) p. 227.

How Do Japanese Aerospace Firms Compare with Those in Other Countries?

Six firms dominate Japanese aerospace activity; namely: Mitsubishi Heavy Industries (MHI), Kawasaki Heavy Industries (KHI) and Fuji Heavy Industries (FHI) which are airframe constructors; Ishikawajima Heavy Industries (IHI) concentrates on aero-engines, leaving Shin Meiwa and Nippi to derive most of their aerospace turnover from parts and components. With the exception of Nippi, all these firms are diversified enterprises having relatively insignificant aerospace divisions (see Table 9.5).

The main companies involved in the aerospace industry – FHI, IHI, KHI and MHI – are predominantly shipbuilders which are under pressure to diversify from a sector where they are losing their cost competitiveness and into growth industries where the need for a high level of technology gives them a leading edge over emerging Asian competitors. Thus Japanese

Table 9.3   Turnover in the aerospace sector in Japan, EC12 and USA

Country	Aerospace turnover (millions ECUs constant 1985 prices)			Growth 1979–89 (%)
	1978	*1984*	*1989*	
Japan	2 399	3 826	5 404	125
Belgium	263	305	311	18
France	7 036	10 668	11 867	69
FRG	3 577	4 639	8 424	136
Italy	1 413	2 423	2 976	111
Netherlands	460	597	1 079	135
Spain	182	390	600	230
UK	7 370	10 473	12 711	73
EC12	20 302	29 497	37 968	87
Canada	2 349	3 205	5 611	139
USA	77 774	85 406	117 239	51

Source: Commission of the European Communities (1992), p. 183.

Table 9.4   Division of aerospace activities by subsector and destination for Japan, EC and USA

	Breakdown of overall aerospace turnover, 1989		
	Japan (%)	EC (%)	USA (%)
By subsector:			
Aircraft and missiles	40	49	49
Space	28	6	10
Engines	14	18	20
Equipment	18	27	21
By destination:			
Civil	45	44	34
Military	55	56	66

Source: Commission of the European Communities (1992), pp. 187–8.

Table 9.5   Turnover and workforce of Japanese and other selected aerospace manufacturers, 1982–90

Company	Turnover (ECU mn, current prices)				Workforce			
	1982	1986	1990	Growth 1982–90 (%)	1982	1986	1990	Growth 1982–90 (%)
Mitsubishi	599	1 648	2 221	271	6 360	6 200	6 500	2
Kawasaki	412	800	1 239	201	3 700	3 900	4 900	32
IHI	331	640	804	143	3 550	3 200	3 550	0
Fuji	105	234	314	199	1 974	2 179	2 572	30
Boeing	9 222	16 708	21 670	135	90 100	118 500	161 700	80
McDonnell	7 073	11 708	11 308	60	60 970	92 280	108 800	78
B Aerospace	3 663	4 671	8 653	136	78 990	75 480	81 869	4
DASA	3 481	3 766	6 204	78	53 233	52 178	58 664	10
Aerospatiale	3 331	3 737	4 741	42	36 450	34 246	33 506	−8
Rolls-Royce	2 664	2 683	5 141	93	48 800	41 900	65 900	35
Dassault	1 967	2 294	2 476	26	15 782	15 783	12 390	22
Alenia	908	2 028	2 743	202	18 827	27 956	30 179	60
Fokker	605	584	1 385	129	8 398	10 860	13 561	62
Agusta	435	580	722	66	10 421	9 703	9 300	−11
CASA	308	337	764	148	9 622	10 591	9 544	−1

Source: Commission of the European Communities (1992), pp. 239–40.

aerospace producers differ from their European and American rivals in that they are not aerospace specialists like Boeing or Airbus.

Data on the turnover and workforce attributable to aerospace activities for the leading Japanese companies are reported in Table 9.5; comparable data for other selected aerospace companies are also presented. The smaller size and faster growth of the Japanese firms reflects the same characteristics for the industry as a whole (see Table 9.3). Mitsubishi, the largest of the Japanese firms, has a turnover one-half of the size of Aerospatiale and one-quarter of the size of British Aerospace.

Following the outbreak of peace in Europe and reduced defence budgets, concern is being expressed about the extent to which some major defence manufacturers are dependent on arms sales for their survival. Table 9.6 compares the defence dependence of the major Japanese defence manufacturers with that of the six largest arms-producing firms in the West. Compared

Table 9.6   Dependence on military sales

Rank	Company	Country	Arms sales ($mn)	Total sales ($mn)	Arms as a per cent of all sales	Employment (nos)
1	McDonnell Douglas	USA	8 500	14 581	58	128 000
2	General Dynamics	USA	8 400	10 053	84	103 000
3	Lockheed	USA	7 350	9 932	74	82 500
4	British Aerospace	UK	6 300	14 898	42	125 600
5	General Electric	USA	6 250	54 574	11	292 000
6	General Motors	USA	5 500	126 932	4	775 000
20	Mitsubishi HI	Japan	2 640	15 180	17	43 914
39	Kawasaki HI	Japan	1 270	5 892	22	16 833
55	Mitsubishi Elec	Japan	810	17 308	5	47 607
74	NEC	Japan	510	20 011	3	38 013
76	Toshiba	Japan	500	22 187	2	69 643
81	Ishikawajima-Harima	Japan	460	4 587	10	26 178

Note: All Japanese companies in the 100 largest arms-producing firms in the OECD and Third World (excluding those in China and the former Warsaw Pact), 1989.

Source: SIPRI (1991), pp. 311–16.

with the top four, the Japanese defence companies have a relatively small proportion of their total sales attributable to arms. The three Japanese companies mentioned for the first time in Table 9.6 are Mitsubishi Electric, NEC and Toshiba. These are largely electronics companies and the nation's main satellite producers.

Is Japan Self-Sufficient in Aerospace?

Japan's specialisation on licensed manufacture or co-production in order to acquire US technology has tended to overshadow indigenous design. However, Japan has produced some indigenously designed aircraft and these are detailed in Table 9.7. The absence of any military aircraft exports is attributable to the self-imposed ban on the export of military technology.

The designs of Mitsubishi in the commuter and business jet fields have been effectively abandoned with the sale of Mitsubishi's US subsidiary to Beech. Consequently, Japan pins most

Table 9.7 Japanese aircraft production of indigenous designs, 1945–90

Aircraft	Firms	First flight	Description	Sales		
				Domestic	Exports	Total
MU-300	Mitsubishi	1978	8-seat business jet	n/a	n/a	106
MU-2	Mitsubishi	1963	11-seat business jet	57	698	755
YS-11	NAMC	1968	60-seat airliner	113	69	182
F-1	Mitsubishi	1975	Jet fighter	77	0	77
C-1A	Kawasaki	1970	Military jet transport	31	0	31
T-4	Kawasaki	1985	Jet trainer	91	0	91
T-2	Mitsubishi	1971	Jet trainer	92	0	92
T-1	Fuji	1958	Jet trainer	62	0	62

Source: *Jane's All The World's Aircraft* (Jane's Publishing Group, Coulsdon, Surrey) (various issues).

of its hopes in civil aircraft on its relationship with Boeing for whom it has become an important supplier of components.

Military projects include the on-going licensed production of the F-15 Eagle and the development of a derivative of the General Dynamics F-16C Fighting Falcon under the programme title FSX. The latter will be developed jointly by General Dynamics and a Japanese industry team headed by Mitsubishi as prime contractor. Japan will be totally responsible for the FSX programme, including all funding, and will sub-contract approximately 40 per cent of the work to US companies.

Table 9.8 provides details of the current major aircraft and aero-engine programmes in Japan. They comprise a mixture of indigenous and collaborative projects, together with licensed production and sub-contract work, largely for US firms. Of the on-going indigenous projects, the Asuka commercial STOL aircraft and the T-4 military jet trainer are the most important. Collaborative projects include a number of Euro-Japanese ventures such as the BK 117 helicopter and the V2500 turbofan engine. Licensed production has been Japan's most favoured approach to the creation and development of an aerospace industry. This has involved the production of many US airframe and engine designs, particularly for helicopters. The only current Euro-Japanese connection comprises a risk-sharing partnership in Rolls-Royce's new large civil jet engine programme. Sub-contract work is again a largely United States–Japan affair although KHI has won a substantial contract from Airbus for

Table 9.8   Ongoing Japanese aerospace programmes

Firm	Project	Description
(a) Indigenous designs:		
FHI	KM-2Kai	Turboprop version of licensed produced piston-engined KM-2 (Beechcraft Mentor) for JMSDF.
IHI	F3	Development of this small turbofan began 1976; XF3-30 selected in 1982 for T-4 trainer.
IHI	ITS 90	Small turboshaft engine under development.
KHI	Asuka	Development programme for a 150-seat commercial STOL passenger transport aircraft.
KHI	T-4	Indigenous design; plan to procure 200 of this jet trainer.
KHI	Unnamed	Small turboshaft engine under development.
MHI	Unnamed	Small turboshaft engine under development.
Shin Meiwa	US-1A	Amphibious search and rescue aircraft.
(b) Collaborative designs:		
Ishida	TW-68	Tilt-wing shuttle aircraft for corporate, commuter, freight and SAR. Entirely Japanese-funded up to first flight of a prototype (due mid-1993); design by an American company whose workforce includes a number of former Bell employees.
KHI	BK117	Collaborative helicopter venture with MBB of Germany. Production continues; sales exceed 180.
MHI	F-SX	Development of derivative of the F-16C continues.
Fuji, KHI, MHI	777	Collaborating with Boeing on development of a new wide-bodied twin engine passenger jet. Japanese companies will develop and produce 20 per cent of the aircraft's parts on a sub-contractor basis, similar to the firms' work on the Boeing 767.
IHI, KHI, MHI	V2500	Partners with Pratt & Whitney, General Electric, Rolls-Royce and SNECMA in this international venture to develop a new turbofan.
IHI KHI MHI	Unnamed	Partners with V2500 participants to develop a prototype engine that will fly at Mach 5, about 6400km an hour. Funded by MITI to the value of Y28bn (= ECU150mn).
(c) Licensed production:		
Fuji	UH1-H	Licensed production of the Bell 205.
Fuji	AH-1S	Licensed production of the Bell 209.

IHI	T58	Licensed production of GE's turboshaft engine.
IHI	T64	Licensed production of GE's turboprop engine.
IHI	T56	Licensed production of GE's turboprop engine.
IHI	F100	Prime contractor for production of this engine for the F-15J.
IHI	GE90	Risk-sharing partner with an 8 per cent share (about $110m) in General Electric's new large civil jet engine programme.
IHI	Trent	Risk-sharing partner in Rolls-Royce's new large civil jet engine programme.
KHI	P-3C	Licensed production plans suggest 100 to be built for JMSDF.
KHI	107	Boeing sold exclusive rights in the manufacture/sale of this helicopter to Kawasaki; 160 deliveries by March 1990.
KHI	CH-47	Japanese requirement for 54 copies of this helicopter which is to be licensed produced.
KHI	OH-6D	Production of this helicopter for JGSDF and JMSDF continues.
KHI	T53	Licence built engine for AH-1S Huey Cobras.
KHI	T55	Licence built engine for CH-47J Chinooks.
MHI	F-15J	Licensed production of F-15 continues.
MHI	SH-60J	Development of S-70 anti-submarine helicopter for JMSDF.
MHI	UH-60J	Licensed assembly of 40 S-70 SAR helicopters for JASDF.
MHI	AIM-7M	Licensed production of Sparrow missile for F-15 aircraft.

(d) Maintenance, refit and sub-contract work:

KHI	A321	Manufactures the fuselage for Airbus.
MHI	F-4EJ	Update (to F-4EJKai) of equipment and weapon system continues.
Nippi	Various	Manufactures components for *inter alia* 767, 757, P-3C, T-4, F-15J, CH-47, V-2500, various missiles and satellites.
Shin Meiwa	Various	Manufactures components for *inter alia* T-4, F-15J, P-3C, MD-11 and 767.

Source: *Jane's All The World's Aircraft 1991–92* (Jane's Publishing Group, Coulsdon, Surrey, 1991) pp. 169–76 and 689–90.

production work on the fuselage for the A321. Clearly there is relatively little Euro-Japanese collaboration at the moment in either the civil or military markets.

The industry, with MITI's support, is concentrating on building strength in two specific areas: short to medium range

Table 9.9   Japanese aerospace trade with the OECD area (current prices)

Year	Exports ($ mn)	Imports ($ mn)	Balance of trade ($ mn)	Imports/ exports
1985	109	1 496	−1 387	13.7
1986	137	1 802	−1 665	13.2
1987	216	1 754	−1 538	7.1
1988	291	2 036	−1 744	6.0
1989	359	1 713	−1 355	3.8

Source: Foreign Trade Statistics, Economic Statistics and National Accounts Division, OECD, Paris.

airliners for regional demand, and high-speed commercial transports. Demand for the former is expected to increase, particularly where traffic is fast growing in the Asia-Pacific region, and feasibility work started in 1989 on a 75-seat commuter plane, the YSX. The YSX will eventually become an international joint project in which MITI is keen for Japanese makers to assume a leading role. Work is also in progress on the development of a high speed commercial transport. The Japanese government has allocated funds for research on prototype engine, materials and airframe design. On this basis it might well be that collaboration with Europe on civil aircraft projects is the most likely future development.

In spite of a substantial array of programmes Japan has a large trade deficit in aerospace products. In 1980, for instance, it imported more than ten times the amount it exported, and its continued reliance on US technology is reflected in the fact that US aerospace exports to Japan in 1983 totalled $1540 million while US imports were only $177 million. More recent aerospace trade data are reported in Table 9.9. This confirms a large balance of trade deficit with the OECD area although the ratio of imports to exports has fallen markedly over the period, suggesting that the performance of the industry is improving rapidly.

The fact that arms exports are prohibited has tended to discourage the development and growth of a Japanese arms industry. This, combined with Japan's desire for latest technology equipment, has made the country one of the leading importers of arms, particularly from the United States. As Table 9.10 illus-

Table 9.10   The Leading importers of major conventional weapons, 1986–90

Country	Weapons imports ($ million, constant (1985) prices)					
	1986	*1987*	*1988*	*1989*	*1990*	*1986–90*
India	3 729	4 582	3 382	3 754	1 541	16 989
Japan	1 780	1 768	2 176	3 163	2 083	10 971
Saudi Arabia	2 413	2 400	2 046	1 427	2 553	10 838
Iraq	2 484	4 440	2 155	1 177	59	10 314
Afghanistan	692	768	1 009	2 183	1 091	5 742
Spain	1 039	1 513	1 580	794	639	5 565

Source: SIPRI (1991), p. 199.

trates, only India has spent more than Japan on weapons imports in the late 1980s.

The Japanese Space Industry

Although the Japanese aircraft industry is relatively weak, the space sector is considerably stronger and this has led some commentators to argue that Japan should forget about aircraft and concentrate on space. Indeed, Japan is about to become self-sufficient in communications satellites, the rockets to launch them and the ground stations to monitor them. With an annual space budget of $1.1 billion, responsibility for the Japanese space programme is divided between ISAS (science) and National Space Development Agency (Japan) (applications). Following four failures between 1966 and 1970, Institute of Space and Astronautical Science (Japan) was the first Japanese agency to place satellites into orbit, making Japan a space nation in 1970. ISAS has launched 19 spacecraft including two Halley's Comet probes and, in 1990, Muses, the first spacecraft to be sent to the Moon since 1976. ISAS is developing an uprated booster for launches in 1995. Costing $135 million to develop, the M5 booster will place 2000 kg into low earth orbit (LEO) and could be used for planetary missions. There are no plans to seek commercial contracts.

NASDA was established in 1969 and has, since 1975, launched 20 applications satellites and experienced one failure.

Its N-I, N-II and H-I launch vehicles are based on the US McDonnell Douglas Delta first stage and Morton Thiokol solid-rocket boosters built under licence in Japan. The final N-II was launched in 1986 and the first of nine planned H-I launches took place in 1987. Launch of the first totally indigenous rocket, the H-II, is scheduled for 1993–4 and this will be sufficiently indigenous to escape American restrictions preventing Japan from using rockets which are substantially dependent on US technology for launching non-government satellites. The H-II rocket is being developed by NASDA with government funds at a cost of $1.75 billion and, although there are no concrete plans to commercialise the H-II, initial speculative forays into the market are being made by a Mitsubishi-based consortium called Rocket Systems. However, orders will be difficult to win. Unlike the mid-1980s, when Arianespace had a monopoly of commercial launch vehicles, competition in this market is now intense and is likely to intensify rather than weaken by the mid-1990s. Competitors include Europe's Ariane, General Dynamics' Atlas, Martin Marietta's Titan and Russia's Proton rockets. With no successful commercial launch record behind it, the H-II is likely to have to offer a price substantially below its rivals to draw custom away from those rockets with a more established reputation.

Japan's reputation is growing in space peripherals, that is equipment used in parts of the industry such as ground support stations. As an example of Japan's success in the ground segment market, 16 of the 20 ground stations built for Eutelsat, which is responsible for the European civil communications satellites, have been supplied by Japan. Only four have been supplied by European companies.[7]

Before evaluating the prospects for further Euro-Japanese collaboration, the next section briefly examines the European aerospace industry and its experience of collaborative programmes.

III THE EUROPEAN AEROSPACE INDUSTRY

Data on the size of the European aerospace industry and constituent firms have been reported above (see Tables 9.2 to 9.6). The aerospace sector in the EC is, in turnover terms, about seven times larger than that in Japan and consequently the

Table 9.11   EC12 civil aerospace trade, 1990 (million ECUs)

	Total			With USA			With Japan		
	Exp.	*Imp.*	*Bal.*	*Exp.*	*Imp.*	*Bal.*	*Exp.*	*Imp.*	*Bal.*
Airframes	8 072	10 064	−1992	2 340	8 134	−5 794	365	13	352
Engines	4 788	3 869	919	3 481	3 363	118	46	2	44
Equipment	662	755	−133	209	637	−428	22	6	16
Other	1 462	1 175	287	310	465	−155	41	90	−49
Total	14 945	15 863	−919	6 340	12 599	−6 259	474	111	363

Note: Exp. = exports; Imp. = imports; Bal. = balance of exports minus imports.

Source: Commission of the European Communities (1992), pp. 170–2.

aerospace activities of Japan's leading manufacturer, MHI, are smaller than those of, for example, British Aerospace or Aerospatiale.

In 1990, the EC12's total trade in civil aerospace products showed a deficit, particularly with the United States. As Table 9.11 reveals, there is relatively little trade between Europe and Japan; what there is shows a small surplus for the EC. Indeed, the relative lack of international aerospace trade between the EC and Japan compared with the extensive trade in other products such as electronics and motor cars suggests the potential for 'opening-up' the aerospace markets between the two groups.

Whereas Japan has produced relatively few indigenous aircraft since 1945, the larger EC member states have developed models across the entire product range from high technology fighter aircraft to large civil airliners. With ever increasing R&D and production costs, however, there has been a trend towards collaborative ventures with nations sharing the costs, risks and benefits flowing from these ventures.[8] Moreover, this move towards joint projects has not been restricted to aircraft. In aerospace a variety of helicopters, engines, missiles and space systems have also been the subject of collaborative development in addition to many other items of defence equipment.[9]

To provide an indication of the number and range of collaborative ventures involving European firms, Table 9.12 details some of the current projects in aerospace. These show

Table 9.12   Examples of current collaborative aerospace programmes

Designation	Year	Participating nationalities	Type of project
BK 117	1977	FRG/Japan	Multi-purpose helicopter
Airbus 310	1978	France/FRG/Spain/UK	Civil jet transport
AMX	1980	Brazil/Italy	Fighter/strike
Harrier II	1981	UK/USA	V/STOL fighter
EH 101	1981	Italy/UK	Multi-purpose helicopter
ATR 42	1981	France/Italy	Turboprop transport
EFA	1983	FRG/Italy/Spain/UK	Combat aircraft
Airbus 320	1984	France/FRG/Spain/UK	Civil jet transport
Eurocopter	1984	France/FRG	Anti-tank helicopter
ATR 72	1985	France/Italy	Turboprop transport
MPC 75	1985	China/FRG	Turbofan transport
NH90	1985	France/FRG/Italy/Netherlands	NATO helicopter
Airbus 340	1987	France/FRG/Spain/UK	Civil jet transport
V2500	1983	Italy/Japan/UK/US	Family of engines
MTR 390	1986	France/FRG/US	Engine for Eurocopter
EJ 2000	1986	FRG/Italy/Spain/UK	Engine for EFA
TRIGAT	1987	France/FRG/UK	Anti-tank missile
ANS	1987	France/FRG	Anti-ship missile

Sources: *Jane's All The World's Aircraft 1991–92*, pp. 106–44, and 681–6; *Jane's Weapons Systems 1988–89*, pp. 142–3, 451 (Jane's Publishing Group, Coulsdon, Surrey, 1988).

that collaborative aircraft programmes have been largely intra-European ventures with a few involving partners from other countries. In marked contrast to the pattern of Japanese–US collaboration, there have been relatively few trans-Atlantic projects.

There are two major on-going collaborative projects with a Euro-Japanese connection, namely the MBB/Kawasaki BK 117 multi-purpose helicopter and the V2500 engine being developed by an international consortium comprising firms from France, Japan, the UK and the US. One hypothesis for the dearth of Euro-Japanese collaboration would be that the few such ventures have proved unsuccessful. However, this has not been the case with the BK 117. Table 9.13 reveals that this project has been reasonably successful, at least in terms of the level of sales achieved and export market penetration. More obviously, the relative absence of Euro-Japanese collaboration reflects the 1954 Mutual Defense Assistance Agreement between Japan and the

Table 9.13   The performance of the BK 117 helicopter and its competitors

Helicopter	Weight (kg)	Range (km)	Seats	Year	Price (ECUs mn)	Domestic sales	Export sales	Total sales	Export total (%)
BK 117	2800	545	11	1983	1.3	40	182	222	82
AS 355	2540	695	6	1981	0.6	110	400	510	78
Bell 222	3470	644	9	1980	1.0	100	88	188	47
Agusta 109	2600	625	8	1975	0.9	109	405	514	79

Sources: Calculated from *Jane's All The World's Aircraft 1991–92*; Commission of the European Communities (1992).

United States. Under this agreement, the United States granted Japan licences for the joint development or production of American defence equipment.[10] European suppliers might perceive this agreement as a barrier to their entry of the Japanese defence market.

The European Space Agency

Another major collaborative venture in Europe is the European Space Agency (ESA). This commenced operation in 1975 and has 13 member states. ESA's activities are divided into two classes, mandatory and optional. Mandatory activities are those carried out under the general budget (for example, administration and maintenance costs of the agency's technical installations) and the science programme budget. Member states contribute to them on the basis of their average national income over the three latest years for which statistics are available.

Optional programmes are those where member states can choose whether to participate and the degree of their involvement. The financial contributions of member states will reflect their interest in a particular field of activity. For example, France has always been the most enthusiastic supporter of an independent European launcher capability and has accordingly been the main contributor to the Ariane programme.

ESA projects include Columbus, a laboratory in space which will be Europe's contribution to the United States-led international space station Freedom; Hermes, a manned space-plane which would ferry people and materials to and from the space station; and Ariane-5, a more powerful version of the Ariane-4 rocket used by Arianespace to place satellites into Earth orbit.

Member states contribute to the mandatory general and science budgets according to their national income, and contribute to the optional programmes according to their level of interest in each particular project. Work is allocated so that in the long run each member should receive contracts in proportion to their funding of ESA's operations (that is, that the overall return coefficient should take the value of unity for each member state).

ESA's budget for 1990 was about \$3 billion, three times the size of NASDA's. However, compared with the US space budget of about \$30 billion, both the European and Japanese budgets are relatively small! Both Europe and Japan are undertaking similar programmes in at least two areas of space research and both appear to be potential candidates for Euro-Japanese collaboration. Cost escalation on the Hermes space plane project, designed to ferry people and materials to the US space station Freedom, has led ESA to seek new partners. NASDA is also contributing to the United States-led programme with an unmanned space vehicle. Euro-Japanese collaboration could avoid unnecessary duplication on these very costly projects. At the same time both parties are developing a new launch vehicle. ESA is pressing ahead with the Ariane-5 and NASDA is developing the H-II rocket. Again, collaboration could offer substantial economic benefits to both sides.

Performance Indicators: Have Collaborative Projects been Successful?

International collaboration offers two major benefits, both of which stem from the economic characteristics of the aerospace industry itself. First, R&D can be extremely costly, and second, production is often subject to substantial learning and scale economies. These mean that unit R&D costs can be a relatively high proportion of total unit costs and that unit production costs depend, to a large extent, on the often related issues of the length and speed of the production run. Collaboration offers the opportunity to ameliorate the effect on costs of both a high R&D burden and short production runs inherent in manufacture for a small domestic market. The partners can share total R&D costs and combine their production orders. Also, for private venture civil aircraft projects, the partner nations can share the risks of the programme.[11] Moreover, the longer production runs generated by an enlarged domestic market are said to reduce unit production costs and thus enhance competitiveness in world markets. But have collaborative ventures been successful?

Numerous problems arise with any assessment of the performance of collaborative projects. With what alternative is the actual joint venture to be compared? For example, should collaborative European ventures be compared with national European ventures or with American projects? Another complication is the necessity of comparing like with like. There are never two identical projects, let alone one a national and the other a collaborative venture. Moreover, how is the performance of any project to be measured? Profitability, one indicator of market efficiency, has a number of drawbacks in the present circumstances and also in profit-controlled defence markets. Accounts for joint ventures are usually submerged within the accounts for the participating firms: hence the identification of costs and revenues attributable to a particular project can rarely be identified. A further complicating factor is that in aerospace there are often hidden subsidies and preferential purchasing arrangements for domestic manufacturers. Nevertheless, several performance indicators can be identified.[12] To examine the performance and international competitiveness of collaborative aircraft projects the focus below is on a combination of output level, market share and export penetration data.

Table 9.14 reports the market share for civil jet aircraft won by the three major producers (Airbus, Boeing and McDonnell Douglas) for the periods 1975–6, 1983 and 1990–1. In the earlier period the market was dominated by Boeing, winning almost two-thirds of all orders, and McDonnell Douglas with just under one-fifth of the market. By 1990, Boeing's market share had fallen to 42 per cent and Airbus's had risen from 8 per cent to 31 per cent, enabling it to overtake McDonnell which retained just under 20 per cent of the market. Clearly Airbus, the collaborative venture, has gained at the expense of Boeing and, to a lesser degree, at the expense of national European producers (whose market share fell from 15 per cent in 1975 to 6 per cent in 1990). However, a comprehensive analysis would also require consideration of the extent of government support and subsidies to the three manufacturers.

Table 9.14   Market shares for civil jet aircraft

Producer	civil jet aircraft ordered									
	1975		1976		1983		1990		1991	
	Nos	(%)	Nos	(%)	Nos	(%)	Nos	(%)	Nos	(%)
Airbus	15	8	1	–	10	5	359	31	101	24
Boeing	109	58	170	69	131	59	483	42	252	59
McDonnell	36	19	41	17	44	20	189	17	33	8
Other European	29	15	20	8	31	14	67	6	40	9
All others	–	–	14	6	5	3	49	4	–	–
Total	189	100	246	100	221	100	1 147	100	426	100

Source: Derived from Commission of the European Communities (1992) p. 23.

Further information on the performance of collaborative aircraft projects is reported in Table 9.15. For large civil aircraft, collaborative projects have not yet achieved production runs commensurate with their US rivals although they are longer than those of non-United States national projects. Their export performance is better than their rivals in both the large jet and regional transport markets. In the military market, combat aircraft have been the focus of most collaborative activity. Although collaborative projects have achieved longer production runs than their non-US rivals, their export performance has been less successful than that of their civil counterparts. For other collaborative military projects the small sample sizes make generalisations difficult. However, for light attack, heavy transport and maritime patrol aircraft the evidence suggests that collaborative production runs are longer than those for non-US national projects but have yet to achieve American scales of output. The export performance of these more specialised aircraft has also been mixed.

A recent survey of EC defence firms also found that for R&D, collaborative European projects typically cost between 10 and 25 per cent more than a comparable national venture.[13] However, with only two partners this will lead to savings for each participant which are likely to increase with the number of nations involved. Most respondents also reported that collaboration reduced unit production costs by between 10 and 25 per cent.

Table 9.15   Post-1960 aircraft: production runs and export performance

	Exports (a)	Total sales (b)	Mean sales (c)	Export performance (a)/(b) (%)
Civil: between 20 and 100 seats				
Collaborative (n = 4)	869	1 072	268	81
American (n = 0)	–	–	–	–
Rest of the world (n = 11)	1 710	2 401	218	71
Civil: more than 100 seats				
Collaborative (n = 5)	1 139	1 346	269	85
American (n = 14)	5 561	11 936	853	47
Rest of the world (n = 5)	435	609	122	71
Military: combat				
Collaborative (n = 5)	254	2 000	400	13
American (n = 11)	3 307	11 324	1 029	29
Rest of the world (n = 15)	1 308	3 263	218	40
Military: trainer/light attack				
Collaborative (n = 1)	161	512	512	32
American (n = 3)	505	3 570	1 190	14
Rest of the world (n = 11)	914	2 056	187	45
Military: transport/patrol				
Collaborative (n = 2)	12	330	165	4
American (n = 5)	908	2 357	471	. 39
Rest of the world (n = 2)	0	59	30	0

Note: Exports and total sales are aggregates for the number of projects shown in brackets.

Source: Derived from *Jane's All The World's Aircraft* (various issues).

However, development periods tended to be longer for joint projects varying from a few months to up to three years.

Although the evidence is at times mixed and inconclusive, there is sufficient to warrant the conclusion that collaboration can offer benefits making joint ventures more attractive than national projects. Indeed, joint ventures resemble an international club in which member states join and remain members so long as membership is worthwhile. Whilst there has been little

Euro-Japanese collaboration to date, are there any reasons why this may change in the near future?

IV   PROSPECTS FOR EURO-JAPANESE COLLABORATION

An analysis of the prospects for Euro-Japanese collaboration demands answers to two questions. First, why should Japan want to collaborate? Second, why should Japan wish to collaborate with Europe (rather than the United States) and why might Europe want to collaborate with Japan? In other words, would a Euro-Japanese club be worthwhile?

Japan has to choose whether to continue with its current policy of licensed production of US military aircraft and a limited involvement in the national development of aerospace equipment. Licensed production can lead to higher unit costs compared with buying directly off-the-shelf from the main manufacturer. In general, cost penalties on licensed production can exceed 30 per cent although there is one reported example where Japanese unit costs in the licensed production of the F104 aircraft were 88 per cent of US costs for a comparable aircraft.[14] Although licensed production might be more expensive than a direct import, it provides compensating benefits in maintaining a domestic defence industry, transferring some technology, providing jobs, together with import savings and economising on costly R&D. In contrast, complete independence is costly. However, collaboration is another option which might be attractive to Japan, providing access to high technology development programmes.

Collaboration is likely to be pursued when the net benefits are expected to be greater than those associated with independence. Various factors suggest that collaboration is likely to be the preferred option in the airframe and aero-engine sectors. In terms of technology the Japanese aircraft industry initially fell far behind the West due to a post-Second World War ban on aircraft production, and has been restricted more significantly by the lack of a sizeable domestic market. In the civil market even if all the Japanese airlines bought only Japanese aircraft it would still not support the development of a Japanese aerospace industry. Moreover, in the commercial market place, the risks and high cost of development and production make it essential to secure firm orders in advance of the project go-ahead. Japanese manufacturers, with no history of product reliability in the aircraft industry, have little chance of succeeding on their own.

There are also the standard economic arguments for collaboration rather than independence. The latter option is often more costly than the former. The evidence suggests that collaboration cuts each partner's R&D costs. Buyers, particularly governments, usually favour domestic manufacturers. Because unit production costs are often a function of total output as well as the rate of output, a wider domestic market means that unit costs and hence prices are lower in highly competitive export markets than they would have been with a national project. In this way the probability of market success is increased.

Indeed, the necessity of international collaboration has been recognised by MITI which will only fund joint international projects. Japan's one attempt to break out of dependence took the form of intra-Japanese collaboration on the YS-11 regional airliner. Its costliness and limited sales convinced MITI that a purely Japanese strategy was not viable. Of course, the bar on the export of Japanese military technology to any country except the United States makes collaboration in defence products with a non-US partner very difficult. In dual-use markets, this prohibition can be circumvented by the foreign partner developing the military application. For example, the military variant of the BK 117 was developed and marketed by MBB rather than Kawasaki.

The attraction of collaboration with Japan is that it offers foreign firms market-entry into both the Japanese and Asian markets, although European companies have found Japanese industry's marked propensity to buy American products a major barrier to entry. On government-sponsored space projects Euro-Japanese collaboration would be one of equals. In view of the cost escalation on the Hermes space plane programme, ESA is actively looking for new partners. Both ESA and NASDA are contributing to the United States-led space station Freedom programme. ESA is providing a manned spaceplane (Hermes) and a laboratory (Columbus); Japan is providing an unmanned winged space vehicle. Collaboration between the two groups might avoid unnecessary and costly duplication on the two space vehicles.

To date, the Japanese have largely collaborated with the United States. Indeed, some commentators have argued that there is a bias against European and in favour of US products. For example, the European aerospace industry has a 27 per cent share of world civil and military markets but only an 11 per cent share of Japanese imports, whereas the US share is 87 per cent.[15] Moreover, this preference for the United States over Europe is not confined to aerospace. For example, Japan's overseas direct investment in North America since 1951 has exceeded that in Europe by almost 150 per cent.[16] There are, though, other explanations for the close United States–Japanese technological relationship reflecting developments following the end of the Second World War, such as the 1954 Mutual Defense Assistance Agreement and the United States–Japan trade balance. However, there are a number of reasons why such United States–Japanese links might weaken in the future.

So far, the US market for aerospace and other products has been much larger and hence more attractive than national European markets. However, with the creation of a single European market and the inclusion of the former Eastern bloc countries, Europe will offer an attractive alternative to the US market. The market for aircraft is likely to grow steadily as the former satellites of Moscow replace their ageing Soviet-built fleets. At the same time moves to de-regulate air travel in the EC are likely to lead to lower prices, to an increased demand for air travel and hence to the need for more civil transports. Furthermore, collaboration with the Europeans will also improve Japanese access to the European market should trade disputes arise. Indeed, some commentators see the recent burst of Japanese inward investment (outside aerospace) being motivated by the desire to circumvent any future fortress Europe policy.

Also, US aircraft manufacturers (such as Boeing) are often unwilling to share project leadership. By collaborating with members of the European industry, the Japanese may be able to obtain more of the high technology work they want. For example, on the BK 117 helicopter, Kawasaki and MBB shared project leadership. Particularly for military projects, European governments, with a long history of collaborative programmes, may be more willing than the US government to share development work with Japan. Furthermore, the recent debate in America over whether the government should permit the export of high technologies for the FSX programme may have encouraged the Japanese to start thinking about alternative partners for future high technology military projects. Euro-Japanese collaboration would reduce both partners' dependence on American technology. Moreover, until the 1980s no civil European airframe manufacturer could rival the market success of the US manufacturers, Boeing and McDonnell Douglas. Hence, collaboration with these firms was likely to have been more attractive to the Japanese than with the struggling European manufacturers whose survival was far from assured. However, as Airbus increases its market share and continues to threaten Boeing's dominant position in the industry, then collaborating with Europe is likely to become more attractive.

In some cases, however, the choice of whether to collaborate with Europe *or* with the United States is not relevant. Where a project is particularly costly, collaboration may involve both European *and* American partners. For example, development of a second-generation supersonic airliner would probably cost between $10 billion and $15 billion and would require participants from Japan, Europe and the United States. Thus there are several factors encouraging greater Euro-Japanese collaboration in the airframe and aero-engine market.

In the space sector, however, the picture is rather different. Apart from their parallel contributions to the United States-led space station, the potential for collaboration in this sector looks far more limited than that in aircraft. Two markets need to be distinguished, namely launch rockets and satellites. In the former, European industry has already coalesced into a single body, Arianespace, which has won about two-thirds of the available market to 1995. Three US launch rockets have virtually all of the remainder of this market.

Japan is in the process of developing its own launch rocket, the H-II, which will compete with European, Chinese, Russian and US rockets. Given the intensity of competition the success of this venture is far from assured. The most likely collaborative outcome would involve a reinstatement of the Japanese link with McDonnell Douglas. Through this venture Japan could associate itself with the US firm's record of successful launches and McDonnell could extend its product range, which is currently limited to a payload of about one-half of the capacity of the nascent Japanese rocket. Given the dominant position of the

European rocket, Arianespace is unlikely to be that enthusiastic about collaboration with anyone. However, as with a new supersonic airliner, the more ambitious space exploration programmes might require international collaboration involving Japan, the EC and the United States.

As far as satellites are concerned, US manufacturers are pre-eminent. Europe is characterised by excess capacity. Although much restructuring has already taken place further adjustment seems inevitable before Euro-Japanese collaboration becomes likely. The Western military satellite market is almost exclusively confined to the United States and thus the potential for collaboration in this sector is extremely limited.

Apart from the benefits available from any individual Euro-Japanese joint venture, there are at least three other more general benefits which are likely to flow from collaboration.

First, Europe undoubtedly has a comparative advantage in collaborative aerospace projects. This is hardly surprising given that it has thirty years of experience negotiating, operating, and evaluating them. By collaborating with Europe, Japan would be able to draw on this wealth of experience and learning. At the same time, however, the absence of a long history of Euro-Japanese collaboration means that such ventures need not be constrained by inefficient practices which continue to impair the performance of many European ventures. The main cause of inefficiency is often the way in which project work is allocated. Although various collaborative bodies, such as the IEPG, acknowledge the long-term goal of allocating work on a competitive basis, a history of *juste retour* often makes the achievement of the competitive goal difficult.[17] For maximum efficiency, partners should specialise in those goods and services in which they have a comparative advantage and contracts for project work should be allocated on a competitive fixed price basis. A policy of *juste retour*, where each partner receives work in proportion to its financial contribution, protects inefficient production and fails to reward efficiency, with adverse effects on the performance of collaborative programmes.

Second, Japan has undoubtedly been one of the economic success stories of the postwar period. Collaboration with Japanese industry offers Europe an additional opportunity to come into contact with this success. Some good practices associated with electronics, manufacturing technology and enterprise culture will inevitably spin off, enhancing the competitiveness of European industry.

Finally, and at a very different level, collaboration will increase the economic ties between the EC and Japan. If the joint ventures involve defence equipment, military ties will also be enhanced. The establishment of a commonality of interest has rendered the use of military means to resolve conflict virtually unthinkable in Western Europe. A similar process of economic and military integration involving Japan and Europe may also contribute to a more peaceful world.

V CONCLUSION

Collaboration between Europe and Japan in aerospace programmes offers potential economic benefits to both groups. Europe has the technical knowledge and experience of designing, developing and producing a complete range of advanced military and civil aerospace products and equipment (for example, combat aircraft, missiles, Concorde, Airbus, avionics, engines). It also has a wealth of experience with international collaborative aerospace programmes where the partner nations have shared high technology development and production work. The United States has very little similar experience.

By collaborating with Japan, Europe would obtain access to both Japanese and possibly Asian markets. Similarly, Japan would bring to the partnership a competitive advantage in electronics and manufacturing together with the ability to pay its entry fee into a collaborative club. In return, Japan would gain access to high technology aerospace programmes and would also achieve a degree of independence from the United States. Initially, collaboration in civil aircraft and ambitious space programmes might be the most realistic and politically acceptable option. However, whilst European–Japanese collaboration in aerospace appears attractive, it is recognised that any developments in this field are in the realm of the long run!

Notes and References

1. This research was carried out under the SPSG/ESRC programme on Defence, Science and Technology Policy.
2. HC299 House of Commons Defence Committee, *European Fighter Aircraft* (London: HMSO, 1992) p. xiii.
3. D. Kirkpatrick and P. Pugh, 'Towards the Starship Enterprise', *Aerospace*, May 1983, pp. 16–23.
4. K. Hartley, *The Economics of Defence Policy* (London: Brassey's, 1991) pp. 107–11.
5. S. Kirby and N. Hooper (eds), *The Cost of Peace: Assessing Europe's Security Options* (Chur, Switzerland: Harwood, 1991) pp. 1–7.
6. R. Drifte, *Arms Production in Japan* (London: Westview Press, 1986).
7. H. Vredeling, *Towards a Stronger Europe*, vol. 2 (IEPG, Brussels, 1986) p. 84.
8. K. Hartley and S. Martin, 'International Collaboration in Aerospace', *Science and Public Policy*, vol. 17, no. 3, June 1990, pp. 143–51.
9. S. Martin, 'Economic Collaboration and Economic Security', in Kirby and Hooper (eds), *The Cost of Peace: Assessing Europe's Security Options* (Chur, Switzerland: Harwood, 1991) pp. 141–70.
10. Drifte, *Arms Production in Japan*, p. 10.
11. K. Hartley, *The Economics of Defence Policy*, pp. 143–4.
12. K. Hartley, ibid, pp. 153–8.
13. K. Hartley, *Collaborative Projects*, Working Paper, Centre for Defence Economics, University of York, 1992.
14. G. R. Hall and R. E. Johnson, *Aircraft Production and Procurement Strategy*, Rand, R-450-PR (Santa Monica, 1967) pp. 156–9.
15. I. Rodger, 'Aerospace Pressure on Japan', *Financial Times*, London, 2 November 1990, p. 3.
16. L. Turner, *Industrial Collaboration with Japan* (London: Routledge & Kegan Paul, 1987) p. 21.
17. Vredeling, *Towards a Stronger Europe*, vol. 1, pp. 3–5.

10 Technology Leakage and US–Japan Security Relations

Michael W. Chinworth[1]

I INTRODUCTION

Technology has been an integral element in United States–Japan security relations throughout the postwar period. The United States encouraged generous technology transfers to Japan in the forty years following the end of the Occupation in order to build up Japanese defence capabilities as a bulwark against Soviet expansion. American policy-makers generally were aware of the economic implications of these transfers – namely, the stimulation provided important industrial sectors in Japan – but accepted them as an acceptable tradeoff to achieve larger strategic goals of providing a security balance against the former Soviet Union. In some cases, particularly in the early postwar years, the positive economic benefits were used to justify transfers because it was felt that a Japan that developed economically would be one less susceptible to communist influences and more inclined to cooperate with US security aims in the region. Thus, there were risks associated with US policies – primarily in the form of technology loss to or through Japan – but they were risks Washington was willing to take.

The economic and security situation facing the United States and Japan in the Asia-Pacific region have altered considerably since these policies were formulated. Japan has transformed itself over the past forty years from being an economic dependent of the United States to its foremost economic challenger. The perceived Soviet threat in the region has altered substantially with the breakup of the former Soviet Union, the slow but steady rise of Chinese influence and other regional scenarios such as a potential reunification of the Koreas. These changing circumstances have led to a re-evaluation of United States–Japan security relations and the role of technology in those ties. The FSX debate indicated that significant elements within the United States no longer are willing to accept the economic risks associated with technology transfers, as well as heightened expectations within the US of Japan serving as a source of advanced technologies in addition to being a buyer of US technology. Many of the concerns once associated with the loss of US technology through Japan have changed because of the reduced Soviet threat. Certain concerns, however, will remain constant while shifting in focus.

Throughout the postwar period, technology issues between the United States and Japan have been characterised by occasional suspicions on both sides of the Pacific. Even as it was transferring advanced military technologies to Japan, the United States feared the possibility of their loss to the Soviets. As Japan's economic presence in the world grew, many in the US came to feel that its Asian ally placed its own economic concerns above what the United States saw as the more important security considerations in the Asia-Pacific region. This suspicion was heightened during the Toshiba Machine Co. incident.[2]

For its own part, Japan often has been wary of US motives in insisting on strict enforcement of technology transfers, often with the sense that US conditions attached to such transfers were meant to retard the growth of key Japanese industries such as electronics or aircraft. To Japan, the FSX co-development programme, for example, was evidence of American fears over the growth of Japan's industry and an attempt to stifle the growth of the industry in the name of security issues.[3] It is likely that as economic competition between the two countries rises, these types of issues will become even more prominent in bilateral technology transfer and control policies.

This chapter will examine technology leakage over the course of the United States–Japan security relationship and the ways in which policy concerns may shift because of a changing economic and security climate. Three forms of technology 'leakage' are considered here. The first is the leakage that takes place in the normal course of cooperative defence programmes to potential competitors in the recipient' country. The United States has transferred massive amounts of information, accumulated knowledge and detailed engineering data to Japanese companies through the course of cooperative defence programmes. Most notable is in the area of aircraft programmes, beginning with the F-86 fighter in the 1960s through the current FSX co-development project.[4] These technology transfers have enabled Japanese firms to boost their indigenous capabilities significantly over time; the question is, to what degree? Have these transfers been coordinated and measured, or have they been a more uncontrolled haemorrhaging of US industrial capabilities to firms that could compete with American companies in the future? The question that has nagged US policy-makers, particularly in recent years, is whether such transfers have also aided commercial interests through defence technology spin-offs. Again, the aircraft industry is the case most frequently cited in the United States, where critics of US policies toward Japan feel that programmes such as the F-15 co-production programme have assisted Japan's longer-term goal of developing an autonomous aircraft industry.

US concerns are likely to be increased in this area because of the growing fear of Japan as an economic competitor. However, because of both generous US transfers and indigenous research and development efforts within Japan, Japanese industry and government have far more options and leverage in accessing US technology than before. Thus, while US fear of the Soviet threat in Asia was sufficient leverage for Japan to gain access to military technology, Japan's domestic capabilities may be sufficient to continue leveraging that access. In addition, Japanese buyers have other alternatives available to them that were not present before, especially in the source of European providers and sellers of advanced systems and technology.

This sort of leakage is inevitable in programmes between two highly advanced industrial nations. Japan will benefit from virtually any sort of joint programme from the United States, if only because government and industry practices routinely emphasise the absorption of technology from abroad to facilitate domestic economic growth. Under those circumstances, the issue facing the United States is not whether to halt such transfers, but the degree of risk the US is willing to assume in joint programmes for reciprocal benefits in light of these new alternatives for Japan (namely, increased domestic development and production, acquisition of products and technology from alternative sources such as Europe, or forgoing certain programmes altogether because of their economic and political costs).

The second form of leakage is of defence-related technologies transferred to Japan or developed indigenously that subsequently find their way into unfriendly hands. This includes the recent case – which is elaborated later – of Japan Aviation Electronics, Inc. (JAE), a subsidiary of NEC Corp. that diverted technology from US fighter aircraft produced under licence in Japan to Iran despite the existence of bilateral and multilateral prohibitions to the contrary.

From the perspective of US security interests, this sort of leakage has been of greater concern to policy-makers but it probably has been very limited. Clearly, technology leakage of this traditional sort has taken place, but it is debatable whether it is more extensive or significant than that which has taken place directly between US firms and other countries. US government officials insist that examples of legal diversions of US technology by or through Japan are few and far between, and until the record demonstrates otherwise, this vow must be taken at face value. With the new global order – or disorder – evident today, officials in both countries will remain concerned about these diversions but the focus will shift from the former Soviet Union to so-called renegade states such as Iraq.

The third form of technology leakage is more complex and may prove to be the most important in the future, especially as Japanese technological capabilities grow and dominate certain industrial areas. This relates to the leakage by Japan of dual-use technologies (falling in the grey zones between purely military and civilian technologies) to other countries. An example of such diversions in recent years was the sale of a commercial dry-dock facility exported legally to the Soviet Union that subsequently was utilised by the Soviet navy to enhance its force projection capabilities in the Asia-Pacific region. A more recent and controversial example was the Toshiba Machine Tool Co. case – where the Japanese firm sold advanced milling machines to the Soviet Union that were utilised in producing more advanced submarines. Although many of these dual-use capabilities are still governed by technology control regimes, those systems have undergone scrutiny as the Soviet Union crumbled and the perceived military threat of the newly organised states of the former USSR has declined. There are still appropriate reasons for controlling the transfer of advanced technologies, but it may be necessary for both the United States and Japan to alter their relationship to effectively manage the risks associated with these leakages.[5]

II ECONOMICS AND DETERRENCE: TECHNOLOGY TRANSFERS IN THE COLD WAR ERA

The United States initiated military technology transfers to Japan shortly after the conclusion of the Occupation era in order to stem what it saw as a growing security threat to Japan and the region as a whole by the Soviet Union. A secondary objective was to stimulate the still depressed domestic economy of Japan. Special purchases by the US military, which totalled almost $2 billion during the Korean War period, convinced Japanese industry of the desirability of military spending and technology to the domestic economy.[6] In addition, the military was viewed as a means of regaining key industries, particularly aircraft, which had been destroyed or disbanded through the world war and subsequent Occupation.[7]

The most significant programme of the period was the F-86 fighter aircraft, which re-established the Japanese aircraft industry. While restricted to what US industry specialists might consider simple 'metal bending' the transfers established the aircraft production capabilities of firms such as Mitsubishi Heavy Industries (MHI) and put them back into a business viewed central to national growth and prestige.[8] These cooperative programmes continued through the F-104, F-4 and F-15. Although the degree of local production varied with each, the level of technology transfers in each programme advanced with the sophistication of the systems themselves. The F-15 co-production programme in particular heightened Japanese industry's desires for the indigenous development and production of advanced fighter aircraft.

Aircraft was not the only industry receiving important boosts during this period. The Japanese radio industry, for example,, started essentially with the local production of US military radios by Japanese firms for US forces in Japan. Sufficient technology was absorbed by the Japanese companies to allow indigenous production. Ultimately, of course, Japanese producers came to dominate local and global markets once the industry got under way. Although the entire success of the Japanese industry cannot be attributed to military-related production, it nevertheless gained its foothold through this avenue, leading later generations of US policy-makers to suggest that the same could take place in such fields as aircraft, modern electronics and other industries.[9]

In addition to these basic inputs, Japanese industry benefited from military-related spin-offs, although the extent still is a matter of controversy. Anecdotal evidence indicates, for example, that the brakes of the famous Japanese bullet trains (the *Shinkansen*) benefited in some measure from experience gained in early fighter aircraft programmes. These instances, however, are not well documented and largely occupy the status of folklore in the history of United States—Japan collaborative defence programmes.

Nevertheless, Japanese industry and, to a lesser extent, the government was convinced early on of the importance of military technology transfers from the United States to Japan for both domestic commercial and military applications. It is this latter area that has been particularly important. US transfers have helped elevate the capabilities of Japanese firms to the point where, coupled with indigenous R&D programmes, they have been able to field several domestic alternatives to systems once supplied by the United States. Japanese missile programmes have been particularly successful, with domestic versions of surface-to-air and air-to-air missiles gradually replacing US missiles (even though the latter arguably are more sophisticated).[10]

In the short term, this has resulted in the loss of sales by US companies to the Japan Defense Agency, and the long-term implication is one of increasing autonomy as well as policy options for Japan. In addition, there are those who feel that defence production is being utilised primarily to support key domestic industries until commercial products can be brought on stream to compete with the United States and other foreign producers (aircraft again is the notable example).[11] Concerned about this trend, US officials have insisted on returning to co-development or other programmes with an even split of production and the United States retaining control over the critical technologies embodied in a particular system.

Japanese development and production capabilities have progressed to a point, however, where government and industry have domestic options rather than sole reliance on the United States for advanced technologies and military systems. Furthermore, Japanese firms can rely on European vendors to provide equal or more generous sales and licensing arrangements, providing additional alternatives to traditional US sources for Japanese imported systems and technology. Finally, the global arms glut as a result of reduced defence budgets in most countries coupled with excess capacity among major arms producers has increased the pressures on most American firms to look abroad, finding a buyers' market that further benefits Japanese interests in terms of technology acquisition.

This is not to suggest that the United States is about to be phased out as a supplier to Japan or that domestic and European alternatives exist for all systems previously supplied by American companies. However, given the proliferation of alternatives now available to Japan, it is unlikely that the United States will be able to roll back the clock and insist on its own terms in every military technology transfer situation in order to minimise the leakage of American technology into Japanese systems, either in the commercial or military sectors.[12] The United States set a course in the mid-1950s that now is difficult if not impossible to reverse. This is especially true because, despite the drawbacks perceived in US policy circles of licensing military technology to Japan, US firms have profited enormously from such arrangements, and within the US context of managing its military technology – namely, leaving large discretion to individual firms – it is difficult to conceive of a situation in which the government will be able to overcome corporate interests to clamp down suddenly on forty years of licensing technology to Japan.[13]

Japanese budgets, of course, are constrained as well. There is a debate within Japan that has yet to reach conclusion over the appropriate level of military funding for the country. On the one hand, officials view the decline of Soviet power as a distinct reduction in the credible threat posed to Japan. On the other, uncertainties surrounding the leftovers of the Soviet military – including the continued high quality of forces stationed in Soviet Asia, uncertainties over command of remaining forces, and the possibility of internal strife spilling over into the Asia-Pacific region – along with other security concerns such as the Korean peninsula and the People's Republic of China, lead many Japanese security analysts to suggest that sustaining present spending trends is at the very least a prudent course to follow for the indefinite future. In addition, the gradual withdrawal of the United States along with US demands for greater burden-sharing by Japan in the immediate area all require continued development of indigenous capabilities in the Japanese perspective. This logically would require the continuation of current technology acquisition and development policies as well.

Both the rationale and management of United States–Japan cooperative defence programmes, then, face re-evaluation because of multiple factors. These include: the success of these programmes in enhancing Japanese industrial capabilities as the country itself emerges as a stronger economic competitor to the United States; radical changes facing defence contractors across the globe; the altered and/or uncertain nature of the security threat in the Asia-Pacific region; and, the appropriate direction of current and future defence spending and acquisition policies in light of those changes.

One of the problems facing policy-makers in the United States in re-examining bilateral programmes is the lack of credible data in so many areas of concern, particularly the extent and impact of technology transfers to Japan. Other than a few studies, the United States simply has not been able to monitor the numerous licensing arrangements between the two countries systematically to allow reasoned judgements on their impact on the competitive capabilities of Japanese companies. A 1967 study broke down such transfers in the case of the F-86 and other programmes in detail, a 1982 audit by the US General Accounting Office (GAO) provided insights into the extent of transfers in the F-15 programme, and the Office of Management and Budget (OMB) examined offsets in the Patriot programme with Japan (but subsequently discontinued those efforts when the OMB determined that the Patriot programme no longer fitted its definition of offsets).[14] Those few efforts and others notwithstanding, there has been no sustained attempt by the US government to ascertain the degree of transfers to Japan and their benefits to potential competitors.

US attempts to monitor this type of technology leakage have been characterised by three problems: their short-term focus, their completion in a highly politicised atmosphere, and the lack of a consistent agreement over such issues as the criticality of certain transfers and the abilities of Japanese firms themselves. The first problem – short-term focus – has been endemic to cooperation programmes with Japan. The GAO and OMB reports cited above are examples of this problem. The GAO study examined the F-15 co-production programme in Japan in a snap-shot fashion. The report listed systems and subsystems licensed to Japan as of 1982, but has never been updated since then. The OMB report was more periodic, but the legislation requiring the study lapsed and no survey has been completed since 1990. Furthermore, its limited definitions restricted in-depth studies to a certain degree.

These and other efforts have been burdened by the politics of the moment in many instances, minimising their usefulness to decision-makers in the United States. An example is the F-15 study, which was influenced by the trade disputes between the United States and Japan in the early 1980s. More recently, congressional and executive concerns have shifted toward 'dependency' concerns, thus diverting some of the attention from technology transfers and potential leakages to questions surrounding investment patterns of foreign firms in the United States and the availability of particular parts and components from overseas suppliers. This shift, which has taken place in less than a decade, reflects the short-term, political pressures injected by the US Congress into defence and security issues, and could easily take place again in the near future.

Finally, there remains disagreement over precisely what areas should be the concern of the United States in monitoring potential transfers and leakages to other countries. The fundamental concern governing the executive branch's decisions on technology transfers to Japan in the F-15 programme was the loss of design know-how to Japanese producers. GAO's focus – which was mandated by the US Congress – was directed more toward the production of US components in Japan (GAO's assumption being that production of US components under licence effectively transferred know-how, including design capabilities, to potential competitors). The two branches have not yet been able to resolve this difference in perspective, making it impossible to form a consistent, government-wide policy toward Japan regarding technology transfers. In the FSX debate of the late 1980s, a comparable dispute existed among policy-makers and between the executive and congressional branches over the criticality of design know-how inherent in the F-16, the base aircraft utilised for the Japanese FSX fighter-support aircraft. Furthermore, significant disagreement existed over the potential value of Japanese technology that was to be acquired over the life of the programme, specifically the advanced phased array radar of Mitsubishi Electric Corp. (MELCO) and the co-cured, composite wing structure technology of Mitsubishi Heavy Industries. Authoritative reports by different US government groups came to decidedly different conclusions when analysing those potential flowback contributions to the United States from Japan.[15] Consistent, uniform agreement over the value of Japanese technologies in this and other cases cannot be expected. However, the inability to reach even general consensus in these areas has hampered efforts to ascertain the extent and impact of transfers to Japan as well as the potential value of flowback transfers from Japan to the United States.

There is a final problem that also is one facing the United States more than the bilateral relationship as a whole. Much of the activity in bilateral defence programmes takes place between companies with only indirect government involvement. The US government assumes that for the most part licensing transactions are matters for individual companies to resolve, with Washington's jurisdiction restricted primarily to the question of the potential loss of technology to unfriendly powers. This attitude has shifted somewhat with the FSX debate, but for the most part individual firms are left to their own resources for marketing and licensing advanced technologies to Japanese counterparts. This complicates the task of the government to monitor transfers systematically, even if the government is inclined to do so.

III   LOSS OF US TECHNOLOGY TO UNFRIENDLY POWERS

If the consequences of legitimate and legal transfers are not fully understood because of lack of awareness and data, then one would assume that the extent of illegal diversions of US technology would be even more unclear. Over time, however, it appears that Japanese firms and the government have been very rigorous in preventing the loss of advanced military technology to unfriendly powers. Exceptions do stand out. The most recent is a case involving the alleged export by Japan Aviation Electronics Industry Co. (JAE) of AIM-9 Sidewinder components produced under licence from the United States to Iran in 1988 and 1989, as well as licensed produced navigational avionics of F-4

Phantom fighters between 1984 and 1987. In response to these allegations, the United States suspended licences of current parts and components to the company, which depends entirely upon defence contracts for its livelihood and is the sole producer of many electronics components for Japanese defence systems (including the important FSX project). Japanese authorities fear that this will set back several domestic programmes, including FSX.[16]

Although this has not resulted in heightened tensions between the two countries to the same extent as the Toshiba Machine incident did (see below), it nevertheless has strained an already complicated picture. JAE's alleged actions would violate international and bilateral agreements: an international embargo against arms exports to Iran, and bilateral licensing agreements barring transfers to third countries. Nevertheless, industry in particular suspects the United States of attempting to utilise these security breaches to punish Japanese industrial competitors. In contrast with the earlier Toshiba Machine Co. case, the outcry in the United States over this incident has been limited. One reason may be the well-publicised involvement of numerous US firms in enhancing Iran's military capabilities, including the Sidewinder case involving JAE.[17] The failure of Congress to criticise the Japanese firms involved in this case to the same extent as the Toshiba Machine incident only underscores suspicion in Tokyo that such criticisms are in fact politically motivated.

Instances of this sort have been few in the course of forty years of licensing programmes between the United States and Japan. Most US officials believe that while there is no doubt that information about US systems has passed into unfriendly hands, the number of violations has been limited. For the most part, the record of the Japanese government and individual firms in protecting US technology from falling into the hands of unfriendly powers has been good. (Indeed, the United States has been accused of allowing greater technology flows to potentially hostile countries through its normal business practices, particularly emphasis on short-term profits and liberal attitudes toward licensing technology.)[18]

In addition, although there remain suspicions about Japan's long-term intentions regarding weapons exports, there is no indication that Japanese exports of components or complete systems are sold routinely to other countries, either legally or in violation of international agreements and/or Japanese policies and statutes. Although the issue appears from time to time in both countries, the Japanese government has reiterated its position repeatedly that it will make every effort to minimise these exports today and in the future.[19]

The third form of technology leakage considered here – sales and other transfers of dual-use technologies from Japan to other countries – is of greater concern in both the United States and Japan. The most notable example of this third form of technology leakage was the sale by Toshiba Machine Tool Co. of advanced milling machines that were utilised by the Soviet Union to produce more sophisticated submarine propellers that reduced noise levels (and thus increased detection difficulties). This affair caught the imagination of the American public and Congress and represented an issue that will remain of concern between the countries in the future. The perception in the United States, to overstate the case somewhat, was that Japanese corporate interests would sell virtually any item to any country for a profit without regard to the security implications.

The involvement of other firms reflected the range of products the United States feared could fall into unfriendly hands. Wako Trading Co. acted as a middleman in the sale. Wako's importance came not from its stature among major trading companies. Indeed, just the opposite was true: it was significant because it cast doubts about the true business of hundreds of comparable small, specialised trading firms in Japan, many of which were established specifically to deal with non-market economies. Almost by implication, these firms fell under suspicion. One of Japan's larger trading companies, C. Itoh & Co., was accused of being involved in the transaction as well (although formal charges were never brought against the company). As the incident grew, Toshiba Machine's parent was accused of even more egregious violations, including the construction of facilities in Eastern bloc countries for producing high-grade integrated circuits for possible missile applications.[20]

Accusations against Toshiba failed to result in formal charges, but the political damage and implications in the case proved far more significant in the long run. Comparable accusations have surfaced from time to time since the Toshiba affair – one involved the alleged sale of equipment to Libya for producing chemical weapons (no formal charges resulted in that instance, either). The number of incidents has been limited, but they followed similar patterns: accusations that fanned emotions on both sides of the ocean and with both sides suspecting ulterior motives on the part of the other. They also demonstrated, moreover, US concerns that if Japan figures in future technology leakage, it will be in the area of dual-use technologies. As Japanese technological capabilities increase, and as the line between purely military and commercial technologies blurs, these transactions could assume greater importance.[21]

IV EXPORT CONTROLS AND BILATERAL FRICTION

The problem remaining at the level of bilateral ties, however, is the widely held suspicion noted earlier within the Japanese business community in particular that US concerns over security violations in reality are disguised attempts to retard the competitiveness of Japanese firms that are competing successfully against American companies. Unable to compete on an equal footing, they reason, the United States has resorted to the guise of international and bilateral security to thwart the competitiveness of Japanese firms. Even members of the presidential administration at the time of the Toshiba case cast doubts on the extent of damage done as a result of the illegal sales.[22] A comparable concern exists within the United States that Japanese businesses are willing to jeopardise regional and international security simply to make a profit.

This suspicion underscores the fact that despite more than thirty years of bilateral technology transfer programmes, the United States and Japan have failed to develop fully reciprocal trust in one another and in the motivations of each country in entering into agreements to prevent the leakage of technology, either to commercial interests from defence programmes or in terms of broader, international agreements. With the rise in both countries of increasing emphasis on economic concerns, and thus the possibility of each nation assuming a larger role as the other's primary policy concern, the difficulty in eliminating this suspicion will only increase.[23]

Thus, continued efforts to minimise technology leakage could become a source of bilateral frictions rather than a security issue viewed in a global perspective. Without a common threat perception, it will be impossible to determine the appropriate level and the common objective of technology controls in the future. This gap is a critical one in the United States–Japan relationship because Japan's military research and development strategy emphasises dual-use technology development and co-development with the United States.[24]

V   TECHNOLOGY LEAKAGE AND THE DEMISE OF THE SOVIET UNION

The lack of a common security threat leads to another problem facing the two countries: justifying continued restrictions on transfers of dual-use technologies with the collapse of the Soviet Union and the former Soviet bloc. The reduced threat posed by the nations, however, does not mean that the world is threat-free or that the United States and Japan lack common security concerns. Chemical and nuclear weapons proliferation stands out as the most pressing security issue still requiring controls over technology leakages, particularly of the dual-use sort in which Japan excels. In regional terms, the Middle East and Korean Peninsula are only two of the many areas that pose problems and concerns common to both the United States and Japan. Controls over the flow of advanced technologies to these areas remains in both countries' interest.[25]

VI   THE PROBLEM OF ENFORCEMENT

One of the difficulties facing the United States in continuing to control the flow of technology through to Japan is the dilemma of developing credible enforcement as the two nations become increasingly interdependent in economic and security terms. In the Toshiba case, for example, initial demands that Toshiba products be banned from US markets ultimately were diluted to affect only a few products for a limited period of time. Congress's demand that Toshiba be punished was tempered by the Japanese firm's intense lobbying in the United States and recognition that attempts to punish Toshiba would only harm US economic interests as well (Toshiba has several plants in the United States that would have lost sales as the result of a ban, and many US companies are dependent on Toshiba-supplied parts and components, without which it would be impossible to assemble their own products).[26] That problem is reflected also in the current JAE case, where Japan is appealing against the suspension of US licences to JAE on the grounds that it will only damage US efforts to encourage Japan to assume greater responsibilities in defending its own shores. By suspending licences, it is argued, JAE will be unable to provide the Japan Defense Agency (JDA) with essential products that allow more effective defence of Japan's territory.[27]

The rise of this interdependency points to a problem in enforcing any technology regime designed to prevent leakage to other countries: namely, that any punishment inflicted on Japanese firms has a high potential for damaging the United States' policy and economic interests as well. Designing credible punishments in order to serve as an incentive to uphold current and future rules, then, will be a problem for policy-makers for years to come, particularly as the United States grows more dependent upon Japanese technology in its own defence systems.

VII THE FUTURE

This chapter began by addressing three forms of technology leakage: loss of conventional US military technologies to Japanese commercial and military sectors; loss of US or Japanese military technologies through Japan to third parties; and, transfers of dual-use technologies to third countries. Having identified some of the problems in enforcing continued controls over these technologies, let us examine the means in which credible controls could be established and existing regimes could be adapted to meet future needs.

First, in the area of loss of US military technologies to Japanese commercial and military sectors, it is reasonable to conclude that to a large degree, the extensive transfers that already have taken place from the United States to Japan have made it difficult if not impossible to return to an age when Japan was entirely dependent upon US technology (and thus also having less leverage *vis-à-vis* the United States to negotiate favourable transfer terms). Transfers over the postwar era un-doubtedly have helped build Japanese capabilities in the military sector (although their benefit to commercial sectors remains unclear). The rise in Japanese capabilities will be supplemented by continued domestic development efforts. At least in the short term, Japan's negotiating position *vis-à-vis* the United States in accessing additional technology should benefit from global defence budget declines in the face of excess industrial capacity. Furthermore, it appears that Japan's defence-related R&D budgets will continue rising.[28] While this does not necessarily guarantee growth in domestic capabilities, it certainly will be utilised in order to assure continued access to foreign technology. The prospects for reversing the 'leakage' of US technology to Japan in this area, then, remains limited. Thus, one factor facing US policy and corporate decision-makers in the future probably will be the degree of transfers acceptable to US interests within the context of bilateral security concerns, not whether it can be reversed or ended entirely. Furthermore, truly essential technologies that are central to the economic and security concerns of the United States must be identified in order to minimise the political frictions associated with future technology transfers in both directions.

In the area of loss of military technologies and dual-use technologies to third countries, the focus of transfers could shift from the Soviet Union to so-called renegade states and potential regional conflicts. As concern over the military power of the former Soviet Union declines, attention could shift toward those nations with demonstrated hostile intentions toward the Western world that also have been able to build up conventional and even nuclear military capabilities. As a nation that maintains commercial relationships through its trading companies with these countries, Japan will come under particular scrutiny to assure that its companies do not violate the agreements its government has entered into in terms of transfers of advanced technologies.[29] These same relationships, however, provide an opportunity to monitor developments in these parts of the world and to minimise the spread of technologies that could be used to harm security in various parts of the globe.

Efforts to control the spread of advanced technologies to these countries could be hampered once again by suspicions of the Japanese business community of ulterior motives on the part of the United States.[30] Even though international controls through such organisations as the Coordinating Committee for Multilateral Export Controls (COCOM) are continuing to be loosened, many Japanese firms continue to oppose broad international control systems for fear that efforts to control technology flows to other nations would hamper business opportunities.[31]

The United States and Japan, then, face three fundamental issues in resolving issues surrounding technology leakage. The first is defining those technologies – if any – that are genuinely critical to bilateral and international security. The second is determining the rationale for continuing technology transfers from the United States to Japan, while expanding reciprocal transfers from Japan to the United States, in the context of these critical technologies. The third is balancing competitive concerns while maintaining cooperation in controlling diversions of technology to unfriendly third countries.

The first area may be the most difficult given the lack of resources available to the United States for tracking technology in both countries, but increased knowledge and awareness can begin by expanded presence in Japan by US corporate and government organisations, a trend that already is under way but has been hampered in part by domestic economic and budgetary problems as well as inconsistent interest by political decision-makers. One reason for Japan's successful absorption of US technology has been the systematic dispatch of government and corporate officials to the United States to identify technologies available for licensing in Japan, as well as general monitoring of economic and market conditions in the United States. Until recently, the United States has failed to mount comparable efforts in Japan, leading to a lack of information concerning: advanced technologies; the utilisation of technology exported from the United States to Japan; and the potential diversion of those technologies to third countries.

The US government has taken limited steps to expand its non-military, defence-related presence in Japan, but efforts often have been hampered by budget restrictions and internal, political disputes.[32] While numerous businesses and academic organisations have increased their representation in Japan as well, far more technical specialists are travelling from Japan to the United States than in the other direction. According to Japanese government figures, over 28 000 researchers visited the United

States from Japan in 1985, seven times the number from the United States to Japan. (Almost nine times as many Asian nationals enter Japan annually for research and study as from the United States.)[33] Until that imbalance can be rectified, it is doubtful that the United States will have sufficient and current information on and access to technology trends in Japan.

Stationing additional manpower in Japan is only the first step required to take effective advantage of information involving Japan's technological capabilities. In addition, closer cooperation between US business and government will be required to diffuse this knowledge throughout policy-making circles. This stands in contrast with the more distant stance taken by the US government toward business up to now, but is an essential requirement if information is to get into the hands of those responsible for identifying truly critical technologies in terms of their impact, availability to other countries, and relative competitive implications for the United States *vis-à-vis* Japan.

In one respect, the United States already has identified critical technologies through periodic government reports.[34] The Department of Defense, Department of Commerce and the White House Office of Science and Technology policy all have identified emerging, next-generation and/or critical technologies that are fundamental to both security and economic competitiveness.

The technologies identified in these reports mirror the results of comparable initiatives in Japan, suggesting a high degree of commonality of views in the two countries. A key difficulty facing the United States and Japan, however, is that these technologies have been identified at least in part because of their commercial potential. This is especially true in the case of Japan's studies. Efforts to control the flow of these technologies to third countries, then, will assume distinct competitive implications, sustaining Japanese fears (however warranted) that security exercises are also aimed at retarding Japanese competitiveness. Having established the criticality of various technologies to bilateral and regional security, the United States and Japan must examine their own security ties to determine the need for and appropriate extent of cooperative security programmes in light of the changing regional and global security environment. It is unlikely that the pattern of the past – extensive licences from the United States with marginal or uncertain flowbacks in return – can continue in the future because of the competitive, budgetary and political concerns discussed earlier. Increasingly, it will be incumbent upon Japanese interests to make accessible, at reasonable prices, indigenous technology to the United States as the price of retaining access to technology for importation to Japan. Although this could drive Japan towards consideration of European sources for more of its defence needs, it is a gamble that must be taken in order to justify the continuation of transfers that, despite occasional difficulties, have been beneficial to both countries. Ultimately, the strength of the US technology base coupled with the guarantees provided by the United States–Japan security treaty should be sufficient to stimulate greater reciprocal flows of advanced technologies between the two countries.

The final problem area is in identifying mutually acceptable technology control programmes affecting critical technologies, while balancing the competitive concerns that arise from such regimes. As has been suggested above, it is probable that the United States and Japan will play a significant role in the development and conscious or inadvertent transfer of advanced technologies critical in significant areas such as missile programmes and advanced electronics. Although total control of these technologies is unpractical and perhaps even undesirable, identification of key areas is essential if the two countries are to restrain jointly the spread of technologies to nations with potentially hostile intentions toward the bilateral alliance, Asia-Pacific region or the global order as a whole.

Where multilateral agreements cannot be obtained in these areas, Japan and the United States should not hesitate to identify programmes that can be enforced effectively on a bilateral basis. A suitable vehicle exists for initiating discussions for identifying such technologies and the means of denying them to nations posing potential security threats: the bilateral Systems and Technology Forum, established under the auspices of the United States–Japan security treaty and a vehicle that has been utilised for such efforts as the negotiation of reciprocal military technology transfers between the two countries and Japan's participation in the Strategic Defense Initiative.[35] Although the body could not necessarily influence international trends in these areas, at least it could be utilised to reach agreement on the rationale for controlling technologies as well as the specific areas that should become the focus of bilateral efforts.

Ultimately, this places Japan and the United States in the position of having to decide which course is truly in their best interests in the future: remaining allied with one another in programmes that possibly restrict business opportunities in other countries, or accepting the possibility that the bilateral security relationship no longer is viable. The willingness of both countries to continue controlling critical technologies, especially the dual-use technologies so central to long-term economic competitiveness, will reflect attitudes on both sides of the Pacific and could shape relations in the future.

The fact that Japan has gone along with COCOM and, in general, has enforced policies to prevent the flow of certain technologies to other countries, suggests that the country has, until now, viewed the maintenance of favourable ties with the United States as central to its economic, political and security policies. The Japanese government continually reminds private sector interests of the dangers associated with being identified as arms merchants: the risks posed to Japan's commercial interests by expanding even technology sales to other countries at present is too high to warrant conscious expansion of transfers and/or sales of critical items and technologies to potential security threats. Only a total rupture of bilateral ties would change that perspective.

If the two countries are unable to surmount the suspicions mentioned earlier that have characterised technology relations to date, Japan at some point in the future may in fact decide that parting with the US emphasis on the control of advanced technologies is worth the risk, if only because the United States no longer has the credibility to enforce sanctions against Japan for possible violations. As interdependency grows, so will the difficulty of enforcing technology control regimes that lack widespread support. Developing the consensus necessary to achieve that support, however, could produce regimes that lack effectiveness and credibility themselves.

The United States and Japan, then, could very well be at a crossroads in their technology relations and in their mutual efforts to control the flow of advanced technologies to third countries in the name of enhanced regional and global security. The course the two countries take could affect not only regional security issues, but the very core of the bilateral relationship as well.

Notes and References

1. The author wishes to thank Hiroo Kinoshita, Masashi Nishihara and Masatoshi Shimbo for their comments on earlier drafts of this chapter.
2. Toshiba Machine Co., a subsidiary of electronics giant Toshiba Corp., sold nine- and five-axis milling machines in 1981 that were installed in the Baltic Naval Shipyard in 1983–4 to produce submarine propellers that would generate less noise and hence would be more difficult to detect by Western anti-submarine forces. The Norwegian government-owned arms manufacturer Kongsberg Vaapenfabrikk provided numerical controllers for the machines. Both sales, which collectively totalled $17 million, violated regulations of the Coordinating Committee for Export Controls (COCOM), a non-treaty, international agreement intended to deny advanced military technologies to the former Soviet bloc. Norway, Japan, the United States and more than a dozen other Western governments were members to the agreement. For an extremely critical view of the Toshiba Machine case, see William C. Triplett II, 'Crimes Against the Alliance: The Toshiba–Kongsberg Export Violations', *Policy Review*, Spring 1988, pp. 8ff. For a Japanese perspective, see Research Institute for Peace and Security, *Asian Security 1987–88* (London: Brassey's Defence Publishers, 1987) pp. 163–5.
3. The FSX is a co-development project between Mitsubishi Heavy Industries and General Dynamics, utilising GD's F-16 as a base for a derivative that will include Japanese technology enhancements. The cost to develop six prototype aircraft is estimated at $2 to $3 billion, which will be born entirely by the Japanese government. For a representative, mass-media treatment of the issue, see Saburo Yuki, 'FSX – Why US Fears Japans Technological Gains', *The 21*, July 1989, pp. 14–17. For a Japanese industry view, see Masao Kuno, 'Who Will Make the FS-X?', *Gunji Kenkyu* (Japan Military Review), April 1988, pp. 38–48, especially pp. 39–41: and K. Sono, 'Nihon no Abionikkus Meika' (Japan's Avionics Makers), *Gunji Kenkyu*, April 1988, pp. 60–3. Two recent – but somewhat sensationalist – publications further criticised US positions in the case, focusing on it as an example of continued pressures and frictions in the technological field emanating primarily on the American side: Shinji Ohtsuki and Masaru Honda, *Nichibei FSX Senso* (The US–Japan FSX War) (Tokyo: Ronsosha, 1991); and, Ryuichi Teshima, *Nippon FSX o Ute* (Stop the Japanese FSX) (Tokyo: Shinchosha, 1991).
4. Major cooperative, fixed-wing aircraft programmes between the United States and Japan with sufficient or near total production in Japan over the postwar period include the following: fighter aircraft: North American F-86 Sabre (February 1958–September 1968); the Lockheed F-104 Starfighter (March 1962–March 1985); the McDonnell-Douglas F-4 Phantom (July 1971–present); and, the McDonnell-Douglas F-15 Eagle (July 1980–present); transport: C-130 Hercules (December 1973–present); reconnaissance: Lockheed P-3C (April 1981–present). The Grumman E-2C is purchased through the Foreign Military Sales programme.
5. Exports of many formerly controlled items have recently been liberalised (for instance, in early 1992, 2000 'sensitive' items were taken off the

COCOM list requiring prior government approval before export is allowed – see John Yang, 'US Relaxes Restrictions On Technology', *Washington Post*, no. 141, 24 April 1992, p. c.2), but disagreements remain over the appropriate level of control of others, particularly dual-use items. In the mid-1980s, for example, the US government regulated the export of such common items as popular electronic games because their internal microprocessors often were more advanced and sophisticated than many microprocessors used for defence applications. Those restrictions have been eliminated in recognition of still newer, more advanced technologies readily available throughout the world. Controls over other consumer electronics and telecommunications products have been relaxed among numerous other items. Controls over even many military items have been loosened. One example relates to various systems utilised in the Global Positioning System.

Disputes remain, however, over the extent of controls required for dual-use and other technologies. A recent defence-journal article noted, for instance, that the US government treats batchmixers as sensitive items because although they commonly are utilised by such industries as bakeries for mixing bread dough, they also can be used to mix rocket fuel (Andrew Lawler and David Silverberg, 'Old Hurdles Stymie New World Trade Order', *Defense News*, no. 12, vol. 7, 23 March 1992, pp. 1, 10, 26). The United States and Japan are among a dozen-plus industrialised nations that belong to formal and/or informal groups that are addressing the issue of appropriate measures for the control of items that can be utilised in ballistic missiles, and nuclear, biological and chemical weapons. These organisations include COCOM, the White House Enhanced Proliferation Control Initiative, the Australia Group, the Nuclear Suppliers Group, and the Missile Technology Control Regime.

6. For an examination of the economic impact of defence spending on Japan's economy, see Nihon Heiki Kogyokai (Japan Ordinance Association), *Nihon Heiki Kogyokai Sanjunenshi* (Tokyo: Nihon Heiki Kogyokai, 1982); Keiichi Nagamatsu, *Nihon no Boei Sangyo* (Japan's Defence Industry) (Tokyo: Orion, Japan, 1979); and Thomas R. H. Havens, *Fire Across the Sea: The Vietnam War and Japan 1965–75* (Princeton, NJ: Princeton University Press, 1987).

7. For an example of early examinations of transfers from military to commercial industries, see Daniel L. Spencer, *Military Transfer of Technology*, AFOSR67, March 1967, AD660537.

8. Concerning the F-86 programme, one US executive at the time said bluntly that 'we were paid to put them into business and we gave them everything we had'. See G. R. Hall and R. E. Johnson, 'Transfers of United States Aerospace Technology to Japan', in Raymond Vernon (ed.), *The Technological Factor in International Trade* (New York: Bureau of Economic Research, 1970) p. 317.

9. Clyde Prestowitz, 'Giving Japan a Handout', *Washington Post*, 29 January 1989, p. D-1. Former Secretary of Defense, Frank Carlucci, responded to Prestowitz's criticisms in 'The FSX Project is No Handout to Japan', in a subsequent issue of *Washington Post*.

10. For an examination of Japan's progress in the development and production

of advanced missile systems, see Michael W. Chinworth, 'Industry and Government in Japanese Defence Procurement: The Case of the Patriot Missile System', *Comparative Strategy*, vol. 9, 1990, pp. 195–243.

11. Richard J. Samuels and Benjamin Whipple, 'Defense Production and Industrial Development: The Case of Japanese Aircraft', chapter 7 in Chalmers Johnson *et al.* (eds), *Politics and Productivity: How Japan's Development Strategy Works* (New York: Harper Business, 1989) pp. 275–318.

12. The recently completed agreement governing the sale of US multi-launch rocket systems (MLRS) to the Japan Defense Agency is viewed in Japan as an effort to turn back the clock. Under the terms of the agreement, the US initially will sell to JDA through the Foreign Military Sales (FMS) programme, with Nissan Motor Co. assembling subsequent knockdown units in Japan. Should the JDA decide to purchase additional units, then, of course, the next logical step in this progression is towards local licensed production of US MLRS units. In any co-production programme, exports of complete units followed by local assembly of knockdown units are the two initial steps that have preceded local production. Given the uncertain status of the defence budget within Japan, however, there are doubts whether the MLRS programme will advance to that point.

13. Sales, licensing and co-production of US systems to Japan have been very profitable for American companies. The licensed production of Patriot surface-to-air missile systems in Japan, for example, will bring an estimated $776 million to the Raytheon Co., the prime contractor of the system in the United States. That value represents nearly 30 per cent of the total sale to Japan, yet is attributable primarily to licensing fees and royalties of $315 million. The US government is expected to gather an additional $87.1 million for fuses, $222.8 million for non-recurrent research and development recovery charges, plus $35.0 million in other fees – a total of $344.9 million (source: US Department of the Army). This is for the first-generation Patriot production in Japan alone, and Raytheon can expect additional profits as Japan upgrades its systems. If previous licence production prices are any indication, Patriot will be deployed for about twenty years, requiring software upgrades and various support service throughout that time. These profits are especially significant in light of the fact that Raytheon developed the system completely under government contract using government funding – no independently financed R&D was provided by Raytheon during the programme. The company's pricing strategy for the system may have been developed with potential overseas sales in mind, but for the most part, these figures represent direct contributions to the company's bottom-line profits and could represent better profit rates than what could be obtained through foreign sales.

14. US Government Accounting Office, *US Military Coproduction Programs Assist Japan in Developing its Civil Aircraft Industry*, ID-82–23, 18 March 1982; Office of Management and Budget, *Impact of Offsets in Defense-Related Exports*, December 1985; *Second Annual Report on the Impact of Offsets in Defense-Related Exports*, December 1986; *Impact of Offsets in Defense-Related Exports: A Summary Report of the First Three Annual Reports*, PB88–1 68802, December 1987. Additional reports were filed by OMB in December 1988 and April 1990.

15. US General Accounting Office, 'US–Japan Codevelopment: Review of the FS-X Programme', *GAO Report NSIAD-90-77BR*, February 1990, p. 29; unclassified executive summary of USAF trip report, May 1990, p. 5. GAO criticised the programme and the potential technological contributions to it by the Japanese partners while the USAF summary was more favourably inclined toward the technological levels of Japanese industry.

16. Naoki Usui, 'Japan Protests License Embargo', *Defense News*, vol. 6, no. 51, 23 December 1991, pp. 1, 20.

17. Michael T. Klare, 'Secret Operatives, Clandestine Trades: The Thriving Black Market for Weapons', *Bulletin of the Atomic Scientists*, April 1988, pp. 16ff. The Klare article cites, among other illegal arms sales, $10 million in shipments of F-14 components from the United States to Iran. The Bush Administration approved over $60 million in sales of high technology items to Iran during the September 1990–October 1991 period. Most of these sales involved dual-use technologies, however, and ostensibly were not related to military applications. See 'Data Show US Approved $60 Million in Sales to Iran', *Washington Post*, 30 January 1992, p. A-6.

18. National Academy of Sciences, *Balancing the National Interest* (Washington, D.C.: National Academy Press, 1987) pp. 203–12. This reference and Wallerstein (see below) contain good summaries of the origins and intent of various US and international export control regimes.

19. US Congress, Office of Technology Assessment, *Global Arms Trade: Commerce in Advanced Military Technology and Weapons*, OTA-ISC-460 (Washington, D.C.: US Government Printing Office, June 1991) especially chapter 6: 'Japan's Defense Industrial Policy and US–Japan Security Relations', pp. 107–20.

20. Daniel Sneider, 'After the Toshiba Debacle: Japan Inc. Treads Carefully', *Christian Monitor*, 30 March 1988, p. 1; David Silverberg, 'DoD Officials Claim Distortion of Secret Data', *Defense News*, 21 March 1988, p. 3.

21. Limited sanctions were levied by the United States against Toshiba Machine Co. in the incident. For example, the company was forbidden to bid on US government contracts for a period of time. Congressional threats to ban all Toshiba products from the United States failed, although they did disappear temporarily from US commissary shelves for a while. The incident did upset the planned transfer of Toshiba missile seeker technology to the US, although that transaction was in jeopardy anyway for technical and other reasons. As a result of the dispute, Japan tightened its control over the export of critical items and expanded the number of customs inspectors to over one hundred (compared to over six hundred for the US). For additional examples of dual-use exports that led to COCOM-related charges by the Japanese government against domestic companies, see Stephen Kreider Yoder, 'Japan Charges 2 Firms Shipped Goods Illegally', *Wall Street Journal*, 6 April 1988; Rita McWilliams, 'Toyota, Yamaha Sell Iran Vehicles', *Washington Post*, 7 April 1988, p. C-1. For an examination of the post-Toshiba Japanese export control regime, see Richard P. Cassidy, 'Japan's Export Control System and Its Importance to National Security', unpublished paper, 31 May 1989.

22. Elizabeth Wehr, 'Deadlock Over Toshiba Ban Stalls Deal on Export Controls', *Congressional Quarterly Weekly Report*, 13 March 1988, p. 673.

23. For an examination of the interrelationship among science, technology, security and US–Japan bilateral relations, see Harold Brown, 'Crossroads for US–Japan Relations', *Issues in Science and Technology*, Winter 1991–2, pp. 24–7.

24. Japan Defense Agency, *Defense of Japan 1990* (Tokyo: Japan Times Co. Ltd., 1990) pp. 141–3; Japan Defense Agency, *Defense of Japan 1991* Tokyo: Japan Times Co. Ltd., 1991) p. 89.

25. Mitchel B. Wallerstein, 'Controlling Dual-Use Technologies in the New World Order', *Issues in Science and Technology*, Summer 1991, pp. 70–7; Yasuhide Yamanouchi, 'Japan's Security Policy and Arms Control in North East Asia', International Institute for Global Peace Tokyo, *IIGP Policy Paper 60E*, October 1991. The Wallerstein article provides a convenient and concise summary of US and international technology control regimes since the turn of the century. For a more extensive elaboration of the conflict between commercial and security interests in technology control regimes and the potential new roles for such systems, see National Academy of Sciences, National Academy of Engineering, *Finding Common Ground: US Export Controls in a Changed Global Environment* (Washington, D.C.: National Academy Press, 1991).

26. Robert A. Rosenblatt, 'Intense Lobbying Cools US Anger at Toshiba', *Los Angeles Times*, 1 May 1988, p. IV–1.

27. Usui. 'Japan Protests License Embargo'.

28. George Leopold and Naoki Usui, 'Japan Research Funding Targets Missiles, Aircraft', *Defense News*, 30 September 1991. For a summary of Japan's commercial and government R&D expenditure trends, see Jon Choy, '1990 Update on Japanese Research and Development', *JEI Report*, No. 37A, 28 September 1990; Science and Technology Agency, *Summary of White Paper on Science and Technology 1990*, October 1990, p. 29.

29. For example, Japan remains the only major industrial power with diplomatic relations with both Iran and Iraq. Two major trading companies, Mitsui Bussan and Mitsubishi Corp., maintain commercial presences in Iran and Iraq, respectively.

30. It should be noted, of course, that the American business community is equally if not more critical of international technology control regimes. One reason, however, is that US firms feel they are hampered in their efforts to sell to other nations while companies based in Japan and elsewhere do not abide by the same restrictions as consistently as those in the US. See Evelyn Richards, 'Confronting the Technology Trade Issue', *Washington Post*, 25 August 1991, p. H-1.

31. *Jiji Press*, 28 November 1991.

32. For example, the attempt to place a representative of the Defense Advanced Research Projects Agency (DARPA) was hampered in part by jurisdictional and protocol disputes between the Departments of Defense and State. The former sought an independent office with a senior official answering directly to DoD, while the latter – backed by the US ambassador to Japan – favoured having a more junior figure housed within the US Embassy in Tokyo and in the State Department's chain of command. The anticipated budget for an independent DARPA office was estimated between $1 million and $3 million annually. A compromise resulted in the

approval of three additional DoD technical specialists being housed within the Embassy to monitor Japanese technology and science. The US Air Force has disbursed $10 million to four US universities to support existing or new programmes aimed at American professionals and students for working in and with Japan.

33. Science and Technology Agency, *Kagaku Gijutsu Hakusho* (White Paper on Technology) (Tokyo: Science and Technology Agency, 1987) pp. 72–6.

34. The initiatives include: Department of Defense, *Critical Technologies Plan*, Report to the Committee on Armed Services, United States Congress, 15 March 1990 (updated annually); and National Critical Technologies Panel (White House Office of Science and Technology Policy), *Report of the National Critical Technologies Panel*, March 1991. On the Japanese side, see Ministry of International Trade and Industry's *Nihon no Sentaku* (Tokyo: Tsusho Sangyo Chosakai, 1988); and the annual defence White Papers, *Defense of Japan*, various years.

35. The Systems and Technology Forum was established in 1980. Its members consist of such officials as the US deputy secretary of defence for international programmes and the Japan Defence Agency's equipment bureau director general. See Japan Defense Agency, *Defense of Japan 1991* (Tokyo: Japan Times Co. Ltd., 1991) pp. 64–5, 182, 195; *Anzen Hosho Handobukku* (Security Handbook), November 1989, p. 144; *Boei Handobukku* (Defense Handbook) (Tokyo: Asagumo Publishing Co., 1991) p. 290.

11 Japan's Policy on Arms Exports
Trevor Taylor

I THE FUNDAMENTALS OF POLICY

Arms exports pose particularly difficult problems for Japan because of two prior, fundamental elements of Japanese security policy whose logic tends to generate pressure for Japan to be quite a significant arms supplier. First, Japan believes that all states have the right to self-defence and itself wants to import arms. Hence it has no clear logical basis for denying its arms in principle to others who need to be able to defend themselves. Second, Japan has chosen to build up its own domestic defence industry while having a limited domestic market. As a consequence, severely restricting arms exports means dearer arms for Japan (the increasing fixed costs associated with many items of defence production cannot be spread over a longer production run).

Yet the driving force behind Japanese arms export policy is derived from Japan's wider constitutional commitment to use force only for the direct defence of Japanese territory, 'to shun traditional instruments of power politics such as mutual defence alliances, power projection capabilities and arms exports', and to keep 'its defence effort to the minimum for the defence of its own home territory'.[1] By and large, Japan has chosen not to export arms.

However, the definition of an arms export is not clear-cut and the overall sphere of activity covered by the heading of arms export policy is not restricted to sales of specialised military equipment. While Japan's tightest restrictions are reserved for a rather tightly defined range of 'arms' and 'military technologies' (see Annex), also involved are regulations and practice about the export of dual-use equipment, which has civil or military uses. Japan's policy towards the United States is clearly special and potentially difficult in political terms. Lastly, Japan increasingly addresses what shape the arms export and import policies of other states might take.

An initial feature to note is Japan's formal policy that arms 'within the concept of self-defence' could be exported.[2] In April 1967, as Japan's defence industries were being built up after their destruction at the end of the Second World War, the then Prime Minister Eisaku Sato published the Three Principles on Arms Export. These said that Japan would not supply arms to communist countries, to states subject to United Nations arms embargoes, or to countries involved or likely to be involved in international conflicts.[3] Nine years later, under Prime Minister Takeo Miki, the government reiterated the three principles. In new policy guidelines the government said that the export of arms to other areas would be 'restrained' and that equipment related to arms production would be treated in the same category as arms.[4]

In wording, such an arms export policy is similar to that of Sweden or Switzerland, for instance, but in practice both these states are ready sometimes to export arms commercially. Japan prohibits particularly the export of complete defence systems[5] and its companies do not even attempt to win sales. The data on Japanese arms sales indicate just how little is allowed. Data collected by the Stockholm International Peace Research Institute (SIPRI) indicate that in 'major' systems during all the 1980s Japan only sold to Saudi Arabia the BK 117 helicopter (which Japan does not consider a military helicopter) and coastguard vessels to three Asian states. Their total value was $55 million but Japan contested even this valuation of its arms exports.[6] Using American data, which applies a wider definition of arms than that used by Japan itself or SIPRI, Table 11.1 compares the Japanese position with that of three other states with comparable defence budgets and of other states with restrictive arms export policies. Between 1984 and 1988 Japan's arms exports were half those of Switzerland.

What makes Japanese policy so notable is not the refusal to export but the parallel, almost contradictory aspiration to build up Japan's defence industrial capability.

II   THE CHALLENGE OF DUAL-USE TECHNOLOGY

Japan also restricts the export of the means of manufacturing armaments but does not seek a complete ban. Japan's basic

Table 11.1   Five-year total arms exports of selected states, 1984–8
(in constant 1988 $ million)

	Arms sales total	1988 defence budget
Japan	833	28 770
Britain	7 608	34 680
France	19 451	35 950
Germany	6 265	35 100
Sweden	1 027	4 975
Switzerland	1 512	3 888

Source: US Arms Control and Disarmament Agency, *World Military Expenditures and Arms Transfers, 1989* (Washington, ACDA, October 1990).

regulation refers to 'machine tools and their attachments exclusively for munitions',[7] but today much manufacturing equipment, including computerised numerical controlled (CNC) machines which can cut and shape metal with great precision, can be used equally for civil and military purposes. Dual-use components of weapons systems present a slightly different but related issue. Most obviously in electronics a range of items made in Japan (such as capacitors, resistors and memory chips) can and do find their way into defence as well as civil equipment around the world. Thus, controlling if not preventing Japanese involvement in the proliferation of others' defence industrial capabilities cannot be a straightforward task.

Japan, which is a member of COCOM, requires exporters to obtain export licences from MITI for items which are on the COCOM Industrial List as equipment useful for military purposes, while MITI's own expertise enables it to identify non-Industrial List items which might nevertheless be of military significance. MITI's guidance to industry stresses that association with arms exports can make the Japanese public turn against a company and generate political pressures in the Diet against business and governments.[8] Before issuing licenses, Japan demands considerable detail about the end use of dual-use equipment and is more wary of allowing an export when there is a military purpose involved.[9] Japan also fulfils its obligations as a participant in the Missile Technology Control Regime, the Australia Group and the Non-Proliferation Treaty, each of which seeks to restrict the proliferation of different non-conventional weapons.

Japan appears willing to supply to some customers machine tools, components and other elements which end up serving defence purposes. Japanese machine tools are a common element in the production investment of Western defence plants. Japanese lap-top computers can be found in bigger defence systems and Japan does not prevent foreign armed forces using some of its 'civil' soft-skin automobile products.[10] Given Japan's dominance of the world market in semiconductors,[11] which may be considered more as a commodity than a product in their own right, it is inevitable that many defence electronics systems include Japanese components. Japan has two significant solid motives for being less restrictive about dual-use items than about defence equipment as such. First, if it was to put what would need to be fierce restrictions on the export of dual-use components and machine tools, Japan's industry would suffer seriously in terms of lost orders. Second, if purchasing states (most obviously those in the European Community) were to feel that Japan might deny them such components or indeed machine tools for defence purposes, they would have strong reason to support their own industries so as to re-establish something like self-sufficiency. The resultant industries would also be dual-use and could become unwanted competition for Japanese firms in global civil markets.

But considerable care seems to be taken in Tokyo so that Japan does not contribute inadvertently to the significant diffusion of defence industrial capabilities. The 1987 Toshiba–Kongsberg scandal resulted in controls being significantly tightened. The list of licensable goods was lengthened and more personnel allocated to the licensing/regulation role.[12] Thus, although Japan has been for many years a major supplier of goods to Iraq, Japanese companies have not been implicated in Iraq's effort to acquire ballistic missile, nuclear and chemical capabilities. It is, however, too early for a final judgement since United Nations inspectors had not, at the end of 1991, been able to find and get access to many Iraqi defence manufacturing plants.

III   THE UNITED STATES RELATIONSHIP

Japan has had a close security relationship with the United States since the Second World War and, under the United States–Japan Security Treaty, Japan enjoys an American commitment to come to the aid of Japan without having to make a reciprocal commitment to assist the United States. About 90 per cent of the defence equipment which Japan buys from overseas is American and Japan's defence industries have been built up with the aid of technology and licences transferred from the United States. In general 'Japan and the United States find long-term interests served by free economic and technological exchanges'.[13]

In these circumstances it is not surprising that as Japanese technology advanced, Tokyo would be unable to maintain the purity of a virtual no arms export policy in the face of American demands that it should be allowed access to Japanese technology relevant to defence. Thus the Chief Cabinet Secretary released a statement in January 1983 that

> the Japanese Government has decided to respond positively to the US request for exchange of defence-related technologies and to open a way for the transfer to the US of 'military technologies' (including arms which are necessary to make such transfer effective) as part of the technology exchange with the US mentioned above; such transfer of 'military technologies' will not be subject to the Three Principles on Arms Export . . . the fundamental objective of refraining from aggravating international disputes, which Japan holds as a nation committed to peace and on which the Three Principles are based, will be secured.[14]

Should a substantial amount of defence technology flow in future from Japan to the United States under the 1983 agreement, a delicate situation could arise on US re-exports of that technology. The United States has accepted legally that Japan should be able to veto the re-export of any technology which it supplies to the US. The detailed arrangements for the 1983 agreement, settled in December 1985, included in para. 4.2 the commitment that

The [US] DoD agrees that the military technology, transferred by the JDA or by a Japanese commercial entity, shall
be used only for DoD military purposes that may be further
specified, as necessary, in the separate memorandum for each
technology and shall not be transferred to any person or
organisation not an agent of the recipient of the military
technology nor any third-country government, person or organisation without the prior written approval of the Government
of Japan.

The DoD also agrees that these provisions on authorised
use and retransfer shall apply not only to the transferred
military technology, but also to the changes, modifications or
improvements thereof, and to any hardware or product produced essentially through an application of the transferred
military technology.

The United States can scarcely complain too loudly in public
about such restraints since the US itself insists on approving the
re-export of any defence technology which it supplies to friendly
companies. In 1991 the United States banned efforts by British
Aerospace to sell its BAe 146 passenger jet (which has American
engines) to Iran.[15] Had the partners in the European Fighter
Aircraft (EFA) opted for a radar based on US technology, as
once seemed possible, the US would have asserted a right to
approve all EFA exports.

However, except for Japan, no other state which has supplied
the United States with defence technology has queried the right
of the US to export that technology to wherever Washington sees
fit. It is not hard to imagine the political reaction in the United
States if a company had the approval of the US Administration
and Congress to export a system and Japan then vetoed the sale.
The fact that the United States feels it can properly veto 'European' civil aircraft sales because of the US technology involved
does not mean it would accept a parallel right for Tokyo on sales
of Japanese technology from the US. It is quite conceivable that
before the turn of the century Japan might wish to veto a transfer
from the US to Israel of Japanese-origin technology in an 'American' fighter aircraft. The political consequences of such a veto
could be enormous.

The 1983 agreement covers explicitly military and not dual-
use technology. Although eight years old, not enough has hap-

pened for any final conclusion to be reached about it. In practice there have not been any 'major' technology transfers in its name and by early 1990 only two items had been transferred.[16] US defence prime contractors and leading subcontractors do not yet procure from firms in Japan and there have not yet been major collaborative defence projects between Japan and the United States. Only in the autumn of 1991 was governmental agreement reached for US–Japanese collaboration on ducted rocket motors.[17] The consequences of President Reagan's Strategic Defense Initiative were that research and development was ordered in the United States and little foreign technology was bought. US defence industry has increased its use particularly of Japanese electronic components, most famously in the Patriot missile, but this has served to provoke debate in the United States about the undesirability of relying on Japan, rather than reassuring Washington that Japan is a reliable source of technology. In general, US interest in Japanese defence technology, although somewhat resurgent at the start of 1992,[18] has come intermittently from parts of the Department of Defense rather than consistently from the US government as a whole.[19] US defence industry is not thought to be well informed about what Japan could offer (see Chapter 10 in this volume) and the restrictions on export from the United States of Japanese defence technology could well have moderated American corporate interest in working with Japanese technology.

Japanese companies for their part see their technology as a vital asset which they are not interested in sharing with others. They would prefer to sell the manufactured sub-systems and components.[20]

The FSX project is discussed in Chapter 8 of this volume, but here it is significant that regulation of re-export by the United States of Japanese-supplied technology for this aircraft was a difficult source of disagreement. The US preference was not to be bound by the terms of the more general 1983 agreement and to be able to export FSX-based technology according to US procedures and law. The Japanese asserted their right to control the re-export of their FSX technology, even from the United States, and wanted only to consider export requests from the United States on a case-by-case basis.[21] Although the Japanese were ready to concede on another issue, the US demand for a payment-free transfer of FSX technology to the United States, they were more resilient on the re-export issue. Disagreement persisted into the 1990s.[22] The United States is particularly interested in techniques for wing manufacture and the radar technologies for the FSX.

Overall, US companies will feel unusually tied in the conditions they must accept for taking Japanese technology under the 1983 agreement. The conditions may discourage US companies from taking Japanese military technology, even when the Japanese are willing to supply it.

It is too early to assess the full implications of the 1991 Japanese Aviation Electronics (JAE) scandal, discussed in the previous chapter, on United States–Japanese defence technology relations, but clearly the issue presents important questions.

Not least of these is whether Japanese readiness to depend on US technology has been substantially reduced following the US government's decision to temporarily suspend all US licences granted to JAE. But on the other hand, the affair may highlight the symmetry which Japan is seeking in its relations with the United States. The United States wants to control the export of US technology from Japan, while even on the FSX, Japan wants to be able to control the export of its technology from the US.

IV   JAPAN AND THE INTERNATIONAL REGULATION OF THE ARMS TRADE

The invasion of Kuwait by Iraq in August 1990 and the scale of the consequent coalition preparations to expel Iraq drew global attention to the international arms transfer issue. Iraq's military capabilities had after all been largely supplied from outside, and in particular from the permanent members of the Security Council. Japan was clearly aware that the $13 billion contribution it felt obliged to make to the coalition effort to liberate Iraq was needed because of the massive amounts of military equipment which permanent members of the Security Council had earlier supplied to Iraq. A much weaker Iraq could have been driven from Kuwait at much lower cost.

In fundamental terms, Japan has never argued that all defence manufacturing countries should pursue the same policy as Japan in refusing to export. Indeed a fundamental element justifying the Japanese defence effort under the terms of the Japanese Constitution is that all states have the right to defend themselves, and clearly, if all states practised Japan's arms sales policy, Japan itself would have been unable to obtain the military equipment it needed once it re-established the Defence Forces after the Second World War.

Japan had long felt that the global arms trade should be marked by moderation and caution[23] but the Kuwait crisis was a stimulus for Tokyo to develop and press its own views as to what arms transfer policies should be followed by those actively involved in the defence trade, both buyers and sellers. As former Prime Minister Kaifu put it, 'Iraq's invasion of Kuwait . . . contributed to a clearer understanding of the shape that the arms control and disarmament process should take in the post-Cold War era.'[24] Japan's basic message was to advocate greater restraint, for instance at the May 1991, Japan-sponsored United Nations Conference on Disarmament in Kyoto.

Like the other major industrial states, Japan is a firm supporter of efforts to prevent the proliferation of non-conventional weapons (nuclear, chemical and biological weapons and ballistic missiles). It plays a full part in the multilateral regimes to control the transfer of unconventional weapons technology and would like to see them improved.[25] Unlike Britain, France and the United States, it sees the abolition of nuclear weapons as a desirable and meaningful goal.

On the conventional arms trade, Japan has persistently asserted the stabilising effects of greater transparency, and so was a co-sponsor, with the European Community states, of the United Nations General Assembly Resolution of December 1991 which established an international register of the arms trade in tanks, armoured fighting vehicles, artillery, combat aircraft, attack helicopters, long-range missiles and warships over 850 tons. It seems likely that Japan will contribute substantially to the cost of maintaining the database[26] and will host the first meeting to discuss the details of its establishment. Japan has argued that 'Not only would more open and accurate information contribute to enhanced trust, it would also serve as a solid basis for meaningful discussion of the arms trade.'[27]

Japan has also urged arms exporters and importers in general to exercise restraint, pointing to such ideas as the amount of military capability needed for defence and local balances of power as guidelines to judge the desirability or otherwise of transfers. The then Prime Minister Kaifu argued in May 1991:

> It is imperative that all nations consider establishing or strengthening their own domestic frameworks for self-restraint in the export of conventional arms. I have taken every opportunity to make this point personally to the leaders of the major arms-exporting countries. Supplier countries should manage their arms export policies in such a manner as to prevent the emergence of a purchaser country building a far greater military machine than is necessary for its defense and thereby creating a military imbalance in the region.[28]

This message was of particular relevance for the former Soviet Union, particularly Russia, and China, two entities which Japan has to address in its foreign policy, and which both present Japan with serious problems. There are doubts about the inclination and capacity of Moscow and Beijing to control effectively the export of arms, because of the limitations of their states' administrative machinery. Also, Japan finds both China and Russia difficult countries in that in each case it is hard for Tokyo to lay down specific conditions for cooperation with and help to them.

Japan is also urging arms importers not to import more arms than are necessary for self-defence and, as the Kuwait crisis unfolded, Japan began hinting that economic aid would not be forthcoming to those who imported arms to excess. Japan's firm feeling on this issue was one element behind the G7 adopting a 'Declaration on Conventional Arms Transfers and NBC Non-Proliferation'[29] at the London Economic Summit in July 1991. During 1991 a policy emerged in which Japan's readiness to provide economic help would depend on a broadly defined range of behaviours by would-be recipients, one prominent element of which was their arms importing policies. Basically Japan emerged as saying that it would be most ready to provide aid to those states which did not:

— import arms to excess;
— try to develop weapons of mass destruction;

— spend too much on the military; and
— abuse the human rights and democratic rights of its citizens.

Of course, there is an obvious question of how such a policy could be implemented in practice. How will Japan decide that a state is importing arms to excess? How will it treat a state which does not import arms but which has a poor human rights record, especially if a state (such as China) is important to Japanese foreign policy? Analytically, it is important to distinguish why aid might be withheld. If a state is considered to have imported too many arms or spent too much on defence, aid might not be given because of a wish to punish the importer for creating a dangerous regional imbalance. Alternatively, aid might be held up because it might be judged that the foreign exchange costs of importing arms were doing such damage to the importer's economy that the beneficial effects of any aid would be much outweighed. Holding back aid because of human rights abuses or because of the development of weapons of mass destruction would be more unambiguously 'punishment' actions. Aid donors must decide if they are particularly worried about the political effects in a region of one state's high defence spending, or about the economic effects in the offending state itself.

This author has been told in Japan that the policy will be implemented through 'an Asian process': when aid is being discussed within the Japanese government, these factors will have a real if hidden impact. Certainly the Japanese government feels that the clear assertion of such a policy will make potential aid recipients think twice about their arms importing and other behaviour. Significantly Japan's thinking in this area is parallel with arguments emerging from the World Bank and the International Monetary Fund which says that help should be held back from those which overspend on defence,[30] so 'Asian processes' will have to occur in decision-making fora other than Tokyo.

Clear in Japanese thinking is that arms export restraints need to be of global application. This is of some significance because, in the debates on such restraints which took place among the major powers in late 1990 and 1991, the United States emerged as wanting to concentrate mainly and perhaps exclusively on the Middle East. Politely, Japan made clear its preferences, when the then Foreign Minister Taro Nakayama spoke in June 1991:

I highly value the Middle East arms control initiative announced . . . by President Bush as a courageous attempt to tackle the intricate problem of arms transfers. The initiative calls for the establishment of guidelines for restraint and of a consultation mechanism among the five leading arms exporting countries. Japan wishes to see an early materialisation of the initiative by the five states. At the same time, Japan considers it important that in the future other major arms suppliers will also participate in the new restraint system and that its scope will be expanded globally.[31]

Interestingly in the same speech the Minister seemed to send a further implicit message to the United States, arguing that Washington should not neglect opportunities to resolve the Arab–Israeli problem. The explicit message was much more general, and reflecting Japan's wider view of the driving forces behind the arms trade:

In order to promote arms control and disarmament measures substantially in any region, including the issue of international arms transfers, it is indispensable to resolve political confrontation.[32]

Foreign Minister Takayama, using medical metaphorsy, argued that arms transfer restraints represented 'symptomatic treatment' whereas the settlement of a dispute was an 'eradicative cure'.

V CONCLUDING POINTS

Japan's policy on arms transfers is quite distinctive. It is the only major country in the world which seeks both to manufacture most of its own arms within its borders (often under licence) and yet which normally refuses to export defence equipment. The consequence of this policy, given the limited size of the Japanese Self Defense Forces, is that the Defense Agency has to pay more for defence equipment. Basically opportunities are missed to spread fixed costs over a longer production run which includes exports. This consideration has led some, especially in the United States, to fear that Japan will in the future re-enter the arms market.[33] However, although there has been domestic industrial pressure for exports dating back to the mid-1970s,[34] Japan has remained more or less outside the international arms market.

This chapter has noted where elements in the character of Japan's arms export policy may be changing. First, Japan has had an arms transfer policy since its remilitarisation after the Second World War which it recognises is not suitable for others to adopt. In the light of the Kuwait crisis, however, Japan has become more positive and explicit about the sorts of policies which it would like both importers and exporters to adopt. However, the concepts on which such desired policies are based, such as restraint, defensive needs and regional balances, are by their nature vague and in need of careful interpretation.

Second, Japan may find it harder in future to control the export of technology useful as components in defence systems or as supporting the manufacture of defence equipment. As more and more high technology has dual-use characteristics, controlling its applications will be more difficult. This is a problem for all developed states, which clearly are most worried about the proliferation of the capacity to make non- conventional weapons and have devised several regimes to address the issues. Some in the West would like to see more intrusive measures to check that equipment supplied is being used for agreed purposes.[35] Given Japan's prominence in many fields of high technology including electronic components, new materials and machine tools, this issue is one to which Japan, and in particular MITI, will have to give particular attention.

Third, Japan's relationship with the United States under the 1983 and FSX agreements could be a source of stress, especially if Japan continues to insist on a right to veto the re-export of Japanese technology supplied to the United States, and if the US takes up chances to acquire much more Japanese defence-related technology.

On a global basis, the arms business is becoming more complex and in future Japan cannot expect to avoid involvement in it. Japan's overseas industrial investment is extensive and growing: although Japanese firms have so far not bought explicit defence companies, they have bought firms with defence interests (such as ICL in the UK, a computer firm owned by Fujitsu) and ball-bearing manufacturers. There are questions, but not yet answers, as to whether Tokyo will want to control the export activities in dual-use technologies of the overseas subsidiaries of Japanese firms? Also, even if the Japanese government professed no wish to control such exports and to leave the matter to the state where manufacturing took place, given Japanese culture, would a company allow its overseas subsidiaries to behave in a way which would not be allowed in Japan?

Again in the future, but perhaps not too far away, Japan will have to take a position on the export for military applications of any collaborative high technology projects in which it gets involved, most probably with American firms but perhaps with some Europeans. This issue could arise, for instance, with regard to jet engines. It is increasingly difficult to isolate armaments politically or technologically from other items covered by international trade and investment. This consideration will inevitably have an impact on future Japanese policies towards the trade in defence-related technology.

Annex

In Japanese regulations:

> *military technologies* 'means such technologies as are exclusively concerned with the design, production and use of "arms"'; and
>
> *arms* are equipments 'used by military forces and directly employed in combat'.

In practical terms, the Japanese government defines arms as any of the following items:

1. Firearms and cartridges (including those for emitting light and smoke) as well as parts and accessories thereof (excluding rifle-scopes);
2. Ammunition and equipment for its dropping or launching, as well as parts and accessories thereof;
3. Explosives and jet fuel (limited to that the whole calorific value of which is 13 000 calories or more per gram);
4. Explosive stabilizers;
5. Military vehicles and parts thereof;
6. Military vessels and hulls thereof, as well as parts thereof;
7. Military aircraft, as well as parts and accessories thereof;
8. Antisubmarine nets and anti-torpedo nets as well as buoyant electric cable for sweeping magnetic mines;

9. Military searchlights and control equipment thereof;
10. Bacterial, chemical and radioactive agents for military use, as well as protection for dissemination, protection, detection, or identification thereof.

Source: Office of Technology Assessment, Congress of the United States, *Global Arms Trade: Commerce in Advanced Military Technology and Weapons* (Washington D.C.: OTA, 1991) p. 119.

Notes and References

1. Kiyoshi Araki, *Japan's Policy in the Regional and Global Context*, RIIA Discussion Paper No. 37 London: RIIA, 1991) p. 5.
2. Andrew Pierre, *The Global Politics of Arms Sales* (Princeton: Princeton University Press, 1982) p. 116.
3. Japan's first guidelines, laid down in 1960, banned the exports of goods which could kill or injure. The Three Principles which were announced in 1967 were internal guidelines from 1962. See Ian Anthony, 'Japan', in Ian Anthony (ed.), *Arms Export Regulations* (Oxford: Oxford University Press/ SIPRI, 1991) p. 106.
4. The Defense Agency, in *Defense of Japan 1990* (Tokyo: The Defense Agency) (English translation by *The Japan Times*), p. 183.
5. Office of Technology Assessment, *Global Arms Trade: Commerce in Advanced Military Technology and Weapons* (Washington D.C., OTA, Congress of the United States) p. 109.
6. Anthony, *Arms Export Regulations*, pp. 103 and 109.
7. Ibid, p. 105.
8. Office of Technology Assessment, *Global Arms Trade*, p. 115.
9. Ibid.
10. For the involvement of Japanese vehicles in the Iran–Iraq war and the Chad–Libyan clashes of 1987, see Anthony, *Arms Export Regulations*, p. 105.
11. 'No Winners in the Chip Race', *The Financial Times*, 21 January 1992.
12. Anthony, *Arms Export Regulations*, p. 108.
13. Hiroyuki Kishino, *Managing the Japan–US Alliance in a Rapidly Changing World*, IIGP Policy Paper 73E (Tokyo: International Institute for Global Peace, November 1991) p. 4.
14. In *Defense of Japan 1990*, p. 301.
15. The release of US hostages in Lebanon made it possible that the US would lift its veto in 1992: see 'US Set to Lift Veto on BAe Sales to Teheran', *The Financial Times*, 30 December 1991.
16. 'US, Japan to Negotiate Military R&D Agreement', *Defense News*, 2 April 1990. US ACDA data (see reference in Table 11.1) give no specific figures for Japan but reveal that arms deliveries to the US and Canada from non-NATO developed states as a group reached a recordable level ($100 million) only one year (1987) between 1978 and 1988.
17. A commitment was also made to work for collaboration on ceramics in the engines of ground vehicles and millimetre wave, infra-red sensors: see

'Japan, US to Team Up on Propulsion Research', *Defense News*, 30 September 1991.

18. 'Pentagon Courts Japan', *Jane's Defence Weekly*, 21 December 1991.
19. 'US, Japan to Negotiate . . .', and 'DoD Picks Up the Pace on US–Japan Joint Technology Effort', *Defense News*, 16 April 1990; 'Bush's Tokyo Visit Boosts Cooperation', *Defense News*, 13 January 1991.
20. Office of Technology Assessment, *Global Arms Trade*, p. 115.
21. 'Japanese Back Down in FSX Row', *The Financial Times*, 29 December 1989.
22. *Aviation Week & Space Technology* reported on 29 July 1991 that 'Technology transfer issues plaguing the programme have been worked out', in 'Joint FS-X Team at Work . . .' p. 44.
23. Anthony Sampson, *The Arms Bazaar* (London: Hodder & Stoughton, 1977) p. 324.
24. Address of 27 May 1991 to the UN Conference on Disarmament in Kyoto.
25. Ibid.
26. See Statement by the then Japanese Foreign Minister Taro Nakayama to the UN Conference on Disarmament in Geneva, 6 June 1991, when he said, 'Japan will be ready . . . to contribute in an appropriate manner to the upgrading and expanding of the database system of the United Nations Department for Disarmament Affairs so that the database will be able to cover data on arms transfers as well.'
27. Kaifu, Address of 27 May 1991, note 24.
28. Ibid.
29. Text dated 16 July 1991.
30. 'UN to Penalize 3rd World Spenders', *Jane's Defence Weekly*, 14 December 1991.
31. Nakayama, (see note 26).
32. Ibid.
33. See Margaronis, cited in David G. Haglund (ed.), *The Defense Industrial Base and the West* (London: Routledge, 1989) p. 260.
34. Pierre, *The Global Politics . . .*, p. 117.
35. National Academy of Sciences (Committee on Science, Engineering and Public Policy), *Finding Common Ground: US Export Controls in a Changed Global Environment* (Washington D.C.: National Academy Press, 1991).

12 Japan's Military Renaissance? Prognosis into the 1990s and Beyond

John E. Endicott

I INTRODUCTION

I believe it is very appropriate to begin this chapter with the Japanese saying, *Rainen no koto ieba, oni ga warau*: if we speak about next year, the devil laughs. Given the events that have occurred in the international system in the last several years, it would seem that anyone given the task of looking more than six months into the future had better have a good pact with the devil. I must assure the reader that I have executed no such alliance. I believe, with a little bit of luck and the analysis provided thus far in this volume, that the future role of Japan's military is a story that can be told with a high degree of certainty. (The luck is in the form of leadership returned – to both sides of the Pacific.)

The reader of this book has been exposed to the work of specialists in the field of Japanese security studies from five different nations: Australia, Canada, Japan, the United Kingdom and the United States. In a sense, this is very fortunate, as Japanese defence and military policy has been presented from the perspective of some states extremely close to the issue – Japan, the United States and also Australia – and some who have the luxury of perhaps more uninvolved analysis – the United Kingdom and Canada. In all cases, one of the strengths of this timely review is that discussants have argued their case from the perspective of national assumptions; I will be no exception, but being last, I have an obligation to reflect on and react to some of the arguments that have preceded this chapter, as well as place emphasis on those items, as an American, that I believe most critical. Most certainly, the issue of the continued United

States relationship with Japan will be covered, as it is fundamental to understanding the course of Japan's future. In addition to the question of the keystone relationship between America and Japan is the notion of democracy and its ultimate staying power in Japan. In fact, how is Japan itself changing in a political and societal sense, what is the relationship of these changes on Japan's concept of national defence, and what is the long-term impact on Japan as a world actor by these developments? Having reviewed the national environment, we will deal with the regional relationships that are so very important – increasingly so – and the question of the need for multinational security infrastructures in the Northwest Pacific and in Northeast Asia. Finally, we will turn to the question of a military in renaissance.

II JAPAN AND THE UNITED STATES

Any discussion of the United States–Japan relationship that does not immediately recognise that the latter months of 1991 and the first half of 1992 were not the best of times between America and Japan would be highly suspect. The chapters by Ron Matthews and Sakanaka Tomohisa, which make steady reference to the importance of the relationship, permit this author to provide additional detail, painful to some degree, but nevertheless useful.

From the cancellation of the presidential trip by George Bush to the Pacific–East Asia region in November 1991, until mid-April 1992, there was a series of events that jolted the United States–Japan relationship, one after the other, causing some to opine that not only were we in a process of transition in our relations with Russia, but we were also witness to the creation of a fundamentally different accord with Japan. Unfortunately, the good news of the United States–Japan Summit over 8–9 January 1992, the Tokyo Declaration, that focused on the coming global partnership between Japan and the United States, was lost amid the misunderstandings that arose when a country's chief executive and head of state is relegated to the role of chief trade negotiator.

This event, as alluded to above, came after a major visit by President Bush to the Pacific was postponed on 6 November 1991 as election results from Pennsylvania indicated that the

American President was spending too much time overseas and ought to focus on the issue of re-election. (After all, was it not Disraeli who commented, 'One must be a politician first in order to be a statesman'?) The problem that occurred, however, was that on 5 November 1991, as Bush was leaving for a European trip, he notified the Japanese that he would not be going to the Pacific as part of a Thanksgiving trip. This information was received by the new Prime Minister of Japan, Miyazawa Kiichi, on his very first full day in office. Not the kind of beginning he would have wanted.[1]

To indicate, however, that he was committed to the United States–Japan relationship, President Bush travelled to Pearl Harbor for the special ceremony marking the 50th anniversary of the 7 December attack by Japanese forces on American territory. While the President's presence was masterful and his words those of one who looked to the future and not the past, the nation, as a whole, led by the TV media, certainly followed closely by the newspapers, participated in what can only be described as an orgy of historical excess. The fifty-year point is special, I am sure, in human events, but the coverage on Pearl Harbor coming after a several-year-long attack by revisionists and sensationalists that questioned even the notion of Japan as a democratic state, and the United States–Japanese relationship generally, resulted in a further deterioration of the already negatively charged American–Japanese political atmosphere.

In mid-December 1991, just prior to the 'joyous holiday season', and several weeks before the now rescheduled visit by President Bush to Japan, came the devastating announcement by General Motors that major restructuring within the mammoth corporation would require the elimination of 50 000 jobs. Days later, it was announced that President Bush would be joined by approximately 21 American executives on his trip to Japan in January, and three of these executives, it was added for impact, would be the leaders of the 'Big Three' automobile manufacturers, including Lee Lacocca of Chrysler. The fact that the Secretary of Commerce, Mossbacher, had resigned from his Executive Department position to assume a leadership position in the Bush Re-election Committee, began to impact directly on trade and defence linkages. It was a period when increasingly it was evident that the power of the Department of State to guide and influence the United States–Japanese relationship was being diminished, almost replaced, by a temporary realignment of power within the Bush Administration.

As history only too well records, the United States–Japan summit of 8 and 9 January 1992 focused on 'jobs, jobs, jobs'. The Tokyo Declaration, a joint statement by President Bush and Prime Minister Miyazawa to outline the path to the New World Order, defined global partnership as a pillar of the United States–Japan relationship and reconfirmed the importance of the United States–Japan Security Treaty.[3] However, the international television coverage of the American President becoming ill, and the general commercial focus of the visit, left all with the belief that this summit was not going to be remembered for its significant security accomplishments. In fact, reports were abroad that members of the Bush entourage had labelled the visit a failure before the President headed for home.[4] The American President did not return home looking in command of his foreign affairs portfolio, formerly his strongest suite.

Soon after the Presidential team returned to the United States the Speaker of the Japanese House of Representatives, Yoshio Sakurauchi, in comments made to a group far from Tokyo – on the Japan-Sea side of Honshu – sent the relationship in a tailspin. Not only, as mentioned in this book's introductory chapter did Sakurauchi describe the United States in 'metal-bashing terms, but went further, labelling American workers as 'lazy and illiterate'.[5]

The fact that these comments were made in camera but were still printed did not make any difference to most observers. But to some who follow closely the changing nature of Japanese society, it was just another indication that the notion of Japan as a closed society and not subject to the scathing review of the press should be reassessed. Cosy relationships between politicians and reporters that had existed for so many years and formed the very nature of the press–politician relationship have become casualties of a new democracy and the need for good stories.

It was not a good story; nevertheless it was published worldwide and resulted in one of the most intense American public reactions in years. 'Buy American' campaigns, public destruction of Japanese cars – most often sponsored by American car dealers or union halls – and the reversal of significant contracts or about-to-be completed business deals were reported in the

press and highlighted in the evening news television programmes.

Following on closely, and seen as an attempt to pour oil on
troubled waters, Prime Minister Miyazawa was quoted as also
disparaging of the American work ethic. Actually, his remarks in
the Diet dealt with both Japan and the United States; that both
had problems in getting the more intelligent youth to resist the
temptation to just make money and not enter the manufacturing
fields. However, quick-draw reporters threw his statements into
the fire, and what was meant to be water turned out to be
gasoline.[6]

Attempts by senior diplomats from both Japan and the United
States to dampen the fury that had been unleashed proved not
too successful. US Ambassador to Tokyo, Michael Armacost,
urged both sides to stop 'hurling accusations across the Pacific'.[7]
His main point, that was interesting in the extreme, was that at a
government level, relations were and continued to be very good.
It was at the general public level that the relationship had got
out of hand. The American Ambassador stressed the point that
as interdependence accelerated – and he saw that clearly hap-
pening – increased friction would occur; it would be helpful, he
opined, for the press to report the positive aspects of this interdependence rather than the negative.

The second-ranking Japanese diplomat in Washington followed some two weeks later with a plea for ending the 'angry,
xenophobic rhetoric damaging relations between his country
and the United States'.[8] Hiroshi Hirabayashi, the deputy chief of
mission at the Japanese Embassy in Washington, contended that
the 'vicious cycle' was initiated by the American side, however
(probably referring to members of the Bush Summit Team that
had called the meeting a failure before they had left Tokyo). He
did note that: 'There have been some recent harsh remarks from
Japan that are not only outrageous, but based on misunderstandings and some kind of prejudice. I regret it.'[9]

The reader should not forget, also, that while this was all
going on at one level, the American candidates for the office of
President had also embarked on their six-month series of primary elections, and were on the campaign trail where reasoned
exchange and the highest standards of debate are seldom found.
The aspect of 'Japan bashing' became so grave that the US Civil
Rights Commission issued a report linking some of the rhetoric

of the campaign with an increase in hate crimes against Asian Americans in the United States. The report, in essence, called on the candidates of all political parties to restrain themselves.[10]

Unfortunately, on the Monday following the release of the Civil Rights Commission report, one of the most insensitive statements of the campaign was made by Senator Ernest Hollings of South Carolina. Speaking to a group of workers at a factory in his home state, he recommended that someone 'should draw a mushroom cloud and put underneath it, "Made in America by lazy and illiterate Americans and tested in Japan"'.[11] This comment drew the angry reaction of US Representative Richard Gephardt that the remark was a racial slur, and a comment from the Japanese official government spokesman that it was obscene.

Racism and xenophobia have been observed during this exchange on both sides of the Pacific. I will not go into the instances that have been reported in the press, but one can make a case that both are at work and need to be controlled by intelligent and active leadership. Opportunistic journalism (some by so-called scholars) and TV reporting are not serving the relationship well at this point. At the time of writing (mid-1992), the economy of the United States shows signs of improving and the Democratic Party candidates have been reduced realistically to one, thus the noise level from candidates trying to vie with each other for the voters' attention has reduced somewhat. There has, however, been a marked impact on the relationship from a popular standpoint. A change has occurred. The question remaining is how serious will it be in the long term. Several polls are available that indicate, on both sides of the Pacific, that some damage has been done, but that there is reason for guarded optimism once the elections are over.

In a *Washington Post*–ABC survey of Americans in February 1992, it was reported that 65 per cent of respondents believed that anti-Japanese attitudes had increased.[12] Two months prior to the February poll only 33 per cent of respondents indicated an awareness of increasing anti-Japanese feelings. One polling analyst noted that 'one of the few things President George Bush managed to accomplish during his recent, ill-fated trip to Japan was to increase anti-Japanese sentiment among Americans'.[13] It was his Vice-President, Dan Quayle, who early on in late January tried to return the dialogue to normal with an impassioned plea to '"avoid mindless Japan bashing", because the US–Japanese trade relationship is improving, not getting worse'.[14]

In a poll released by the Foreign Ministry on 14 April 1992 (but taken in early 1991 after the Gulf War), some clear indications of rising American mistrust of Japan are evident.[15] Forty-two per cent of the American general public (general respondents) indicated they 'mistrust' Japan. That figure, in a poll taken annually since 1960, showed that it had reached the highest point since 1960. The good news, however, is that in 1960 the level of mistrust was 55 per cent, and the actual percentage of those indicating that Japan was 'dependable' was 40 per cent, an increase of 5 per cent from 1991. Those who indicated 'mistrust' listed economic competition (10 per cent) and memories of the war (9 per cent) as reasons for their attitudes. However, the area that was reduced by half when compared to the previous year was in an assessment of the status of relations between the United States and Japan. Only 26 per cent replied that conditions were 'good'.

While most indicated that relations had declined somewhat bilaterally, almost 70 per cent of the general public viewed Japan as a 'stabilising force in the wake of the Soviet Union's collapse'.[16] In fact, the number of Americans who responded that Japan had made an 'appropriate contribution to the international community' increased from 35 to 52 per cent. This is an especially important figure when it comes to the issue of Japan's non-military response to the needs of the 'New World Order'.

As far as the Japanese are concerned, in a poll conducted by the Japanese government in October 1991, prior to all the work up to Pearl Harbor and the President's trip, some 78 per cent of respondents indicated that they felt 'close' to the United States. That was a 4 per cent increase over a survey the year prior.[17] Even after the events described above, Yukio Okamoto, a long-time analyst of US–Japanese relations, said that most Japanese still have a fear of divorce from the United States. The concept of 'an isolated child in the world . . . is still the ultimate nightmare for policy planners and ordinary people alike'.[18] However, Professor Honma Nagayo does worry that the Japanese may begin to view their country as 'superior in all respects' to the ailing nation to the east. If this misperception occurs, Honma believes, 'then Japanese–US relations will face catastrophe again'.[19]

A poll that was published in early November 1991 by the

Yomiuri in Japan and Gallup in the United States showed some interesting perceptions of both sides before the President's visit was cancelled. At that time some 41.8 per cent of American respondents thought that trans-Pacific ties were in good condition.[20] That was down, however, from 49.4 per cent in 1990 and from 59.7 per cent in 1986. The November 1991 score was the lowest ever recorded. Some 40.6 per cent of the Japanese, on the other hand, replied that the situation was good. That figure had remained practically constant for some six years. Almost 19 per cent of the Japanese blamed the United States for economic friction, but 28.5 per cent of the Americans blamed Japan.

While the charges and countercharges flew in the early months of 1992, one event that was clearly disturbing was the development of *bubei*, or contempt of America. Unlike the days when the socialists and communists and others were demonstrating against America, the anti-Americanism had good old-fashioned dislike (or *kenbei*) as its main thrust. The newer feelings of the Japanese toward America in the period after President Bush's visit turned more to the *bubei* notion. Inherent in this concept is the idea of superiority. That is disturbing, as one cannot say that things Japanese are universally superior to those American. It is an indication of a world view that plagued the Japanese Imperial Army in the 1930s, resulting from single point analysis and acceptance of bogus doctrines as fact. Certainly, the Japanese do make very good cars. Even if they are superior to every American car built, a belief of general superiority, based on the fact that in some areas Japan has progressed to excellence, can only lead to heartbreak if pursued as a belief system. It is the same kind of 'logic' used to build a morally superior system that resulted in disaster for the Japanese people in the 1940s. The only answer at this point is to reinvigorate the leadership potential both nations have and proceed with the repairing of the relationship. For years this relationship was handled by a few individuals who worked out everything by themselves (usually). One of the prices of democracy is general population involvement in the policy process. It is a price well worth the cost, but it demands a much greater willingness on the part of leaders on both sides of the Pacific to be more proactive to events. Listening to opinion polls and reacting is not leadership. What occurred from November to March 1992 was a mixture of politics, leadership miscues, recession, and loose tongues. The impact on the relationship will in the long run be managed only by the continued development of common interests to the point that they overwhelm and make mute those areas of dissonance. In essence, it appears that both sides have emerged with a new respect for each other; however, they are somewhat battered by the experience. The Japanese have assigned a new Ambassador to the United States, and made other personnel changes throughout America to increase their ability to relate to their most important ally and trading partner. The United States, on the other hand, has let it be known that it, too, will send a new representative to Tokyo, once the 1992 presidential election has given a mandate to one party or the other.

Changing our official 'voices' in each other's capitals, however, is not the answer. It is just a visible indication that this relationship matters. The answer must be found in broad approaches that address wide audiences. A reinvigoration of the 'people-to-people' programmes first started during the Eisenhower Administration would also help. In this era of the popularisation of the policy process, we would be well advised to pay more attention to basic information programmes between the peoples of both nations. The United States–Japanese relationship is basic to a new world cooperative system; at this point, it would be advisable to invest our time and our money to ensure its positive evolution. In this I am confident.

III A CHANGING JAPAN

Up to this point, we have reviewed the dynamic relationship at a national level that exists between the United States and Japan. We have not looked inside Japan, however, to comment on the social and cultural dynamism that exists in that nation and that forms the basis for any military and any renaissance that its military might have. While this is a subject that could involve, and should involve, much greater scholastic coverage, we will introduce only the surface evidences thus far available to the long-time observer.[21]

The readers of this book should know that Japan is a fast changing society. It is not, as some have insisted in revisionist tracts, a society incapable of making decisions (or any less so than the United States), undemocratic in nature, and largely the product of general conspiracies. It is, I would contend, the greatest example of American foreign policy and social management success in the 200-year history of the American Republic. In essence, the US Occupation arrived in Japan with instructions generally to demilitarise the society, democratise it, and to the greatest extent possible, disestablish the *Zaibatsu*. When in the course of the developing Cold War it became clear that a prosperous Japan would make a better bulwark against communist expansionism than an impoverished one, the latter goal was modified somewhat, but not completely forgotten.

Under the leadership of the 'American Caesar',[22] reforms were effected with the cooperation of the Japanese bureaucracy and people that accomplished these goals. Most important, a new constitution was created that incorporated the following features into the revised Japanese political system:

— Authority of government was derived from the people.
— The emperor was made titular head of state.
— War was renounced as a sovereign right of the nation.
— Extensive civil rights were given to all citizens.
— The peerage was abolished.
— Women obtained equal rights with men.
— A parliamentary system of government based on a popularly elected Diet was created.
— Legislative and fiscal authority was given to the Diet.
— Executive power was placed in the hands of a cabinet responsible to the Diet.
— An independent court system was created.
— Local government was decentralised.[23]

These normative goals did not change Japan overnight; in fact, one could say that the reforms incorporated in the current Constitution ensured that Japan became a democratic state in an institutional sense. For more than 45 years the changes inherent in the Constitution were integrated into the Japanese system. Within this system, however, the more traditional cultural and societal organisational norms and concepts were still operative. Thus, the power of the group was seen as more important than the rights of the individual. The concept of seniority as a prime organising concept throughout society was respected. The role of women was limited by educational and reproductive pressures, consumer needs were restrained by producer interests and the overall goal for national reconstruction, and consensus as a decision-making technique made reaching decisions painfully slow. In essence, democracy had been achieved at the institutional level and was guaranteed by the Constitution, but the individual was still operating in a patron–client setting similar to non-industrialised societies.

A series of pressures over the 1980s has placed the old system very much at risk. Change is definitely afoot and moving at an astonishingly fast pace (all things are relative). The kind of change that is occurring involves the place of the individual in society and the relationship of one person to another. Permit me to point to a few indicators of the phenomenon and hope that the sociologists among us who read this will put greater form and substance to simple observations.

In the early 1980s, it became increasingly clear to senior Japanese officials that new technology, always of interest to Japan, was becoming ever more difficult to access. Growing reluctance of American holders of patents to enter into joint projects with Japanese producers was evident as early as the early 1980s.[24] This growing reluctance to share technology at attractive prices put greater emphasis on younger Japanese who were graduates of Western graduate schools, or had PhDs especially from universities in America. Gradually, it became obvious that in the research and development community there was a growing need to stress innovation, creativity and expertise in the race to compete in the international market. Increasingly younger individuals were given R&D responsibilities in laboratories without regard to their seniority, but based on their subject knowledge. They were, in essence, placed in positions of responsibility without regard for seniority.

Sometime in the mid-1980s a new term started to emerge in the Japanese language:[25] it was *shinjinrui*, or new human beings. This word captured what was seen as an evolving new Japanese young person. Generally, it applied to both sexes, and described youth who were increasingly self-oriented, pleasure seeking, 'now' oriented, well educated, somewhat spoiled, with a wider world vision than their parents' generation, and predictable only in that they would do the unexpected.

Such individuals increasingly became difficult to mould into company activities. Rather than drinks at night with fellow workers, they sought out friends, or went home to their wives, if married. When required to go on a company retreat or short vacation to a hot springs in the mountains for a weekend of group relaxation, rather than congregate with workers from the same office they sought out similar-age colleagues and drank together, thinking up new words to call their bosses and members of the more staid generations. Eventually, they started leaving their original companies of employment and sought to use their recently developed expertise to obtain pay increases and a better lifestyle. Asking more about vacation time than retirement benefits when first hired, they made it clear they were not prepared to wait a generation to enjoy life.

This phenomenon, first only among young people, is spreading to the white-collar workers, of whom, recent polls indicate, almost 50 per cent literally fear dropping dead from *karoshi*, or overwork.[26] The notion of 'slowing-down' has received major support from one of the most famous leaders of Japanese business, Morita Akio. In a controversial memorandum, Morita voiced his concern that Japanese workers, generally, were working too hard and being paid too low to 'fully enjoy life'.[27]

The tendency to seek a better job had progressed so far that in 1991 it was announced that approximately 30 per cent of the new employees who had joined major firms in 1988 were seeking new employment.[28] This phenomenon has become so common that many personnel offices in major firms have added a new division: it handles the socialisation of workers making mid-career changes – something relatively unheard of at this level since the postwar era began.

Not missing a chance to capitalise on a good thing, there are now approximately 200 'body snatchers' or firms in business to facilitate personnel intercorporate moves throughout Japan. In Tokyo, some 100 such firms are finding their fortune in helping the new generation move to better-paying and more comfortable surroundings.

Add to this revolutionary assault on the concept of seniority the increased influence and impact of women in the Japanese workforce. Here we are talking about 40 per cent of the workforce in 1992. It is increasingly asking to be heard and it is increasingly the subject of growing respect. OL is a term meaning office ladies, the large number of women who historically have brought the tea and served as administrative support. Now it is increasingly a term that can strike fear into the unwary manager's life. Increasingly demanding career opportunities, not just short-term employment until marriage, the vocal woman's movement received very important backing from the Japanese court system in April 1992 when it upheld Japan's first sexual harassment judgement.[29] The women are also important in that they are adding strength to Japan's growing consumer movement, and 72 of them were admitted in 1992, for the first time, to the new class of the Japanese Defense Academy.[30]

The increasing independence of the Japanese youth is also being reinforced by the relative decline in their numbers as a proportion of the population. Illegal foreign workers are finding opportunities at the lower end of the job spectrum as the Japanese young workers avoid jobs considered dirty, dangerous, or hard. As of February 1992, the Justice Ministry indicated that more than 150 000 illegal foreign workers were in Japan.[31] This, of course, increases the leverage of the youth as a class.

How has this affected national security? The graduating classes of the National Defense Academy for 1989–91 have seen the steady growth in the percentage of graduates who choose not to serve after a four-year free education. The percentage of 'non-serving' graduates became 20 in 1992. Alternative job opportunities are too abundant and there is little or no stigma attached to taking the degree and running.

An early 1992 poll indicates that young men indeed reflect the image mentioned above. It sampled the nation's willingness to put the national welfare above personal gain and came up with some very interesting findings. Only 31 per cent of men in the 25–29 age group would put the nation's well-being above personal gain. Also, the survey indicated that while 64 per cent of respondents generally wanted to be 'useful to society', less than half of those in their 20s felt that way.[32]

What political parties do these young people support? The results of two polls gave some good news and some bad news. In December 1991, a *Yomiuri* poll indicated that 39 per cent of respondents in their twenties supported the Liberal Democratic Party, 9.4 per cent the Japan Socialist Party, 2.7 per cent the Komeito (or Clean Government Party), and 1 per cent each for the Japan Communist Party and the Democratic Socialist Party. Some 45 per cent did not support any party.[33] A little more than one month later, the same newspaper's poll registered a 2 per cent drop in support for the LDP, a 2 per cent increase for the Socialists, and a decrease in all the other parties. Another increase, to 46 per cent, went to the 'none of the above' option.[34] These are additional indications that a significant portion of this age group is not focusing on national issues or ways to influence national issues. One might say that their external political efficacy is quite low, and alienation from the political system high.

As the society becomes more and more able to operate below the level of the group, and as the democratic guarantees of the Constitution become internalised and attainable at the individual level, the traditional norms of historic Japan fade to the fringes of culture and society. Concepts to include *bushido* (the way of the warrior) retreat to the Kabuki theatre stage and certain right-wing political organisations. Society begins generally to open up and the democratisation of it at all levels and in all aspects continues.

This does not make governance easier, and the course of the future is no clearer for having this surge of individualism move throughout the nation. It should assure the world, however, that Japan of the 1990s is not the Japan of the 1930s, and that its military will be of, and does reflect, the society generally. How the military will develop and be used in the regional and international security environment of the 1990s and beyond will be in the final analysis the result of the interaction of external determinants with the domestic setting. The question still unanswered is whether or not all this will constitute a renaissance for the Japanese military; in the sense that we use the word for rebirth, perhaps. However, the nature of the officer corps will reflect the society from which it is drawn. The military of Japan in the coming decade will be prepared to support, in a modest way, the developing international comparative system. It need not be feared, nor should it be underestimated. In the eyes of the serving professionals of the Japanese Self Defense Forces, such a status would, indeed, be a renaissance.

Notes and References

1. *The Daily Japan Digest*, 7 November 1991.
2. This was the unfortunate response to reporters made by Bush when asked why the trip to Asia.
3. *The Japan Times*, 2 February 1992.
4. Ibid, 27 February 1992.
5. Ibid, 4 March 1992.
6. *The Daily Japan Digest*, 4 February 1992.
7. *The Japan Times*, 8 February 1992.
8. Ibid, 27 February 1992.
9. *The Japan Times*, 27 February 1992.
10. Ibid, 7 March 1992.
11. Ibid.
12. Ibid, 20 February 1992.
13. Richard Morin, writing in *The Japan Times*, 20 February 1992.
14. *The Japan Times*, 28 February 1992.
15. Ibid, 15 April 1992.
16. Ibid.
17. *The Japan Times*, 20 February 1992.
18. Ibid.
19. Ibid, 4 March 1992.
20. *The Japan Daily Digest*, 5 November 1991.
21. Please excuse me for assigning that term to myself. I married into Japan in 1959, and have followed the changes to Japan and its people for more than thirty years with great interest. Many of the observations in this section are my own and are based on interactions with an ever increasing network of relatives and friends.
22. With acknowledgement to William Manchester's work on Douglas MacArthur, *American Caesar: Douglas MacArthur, 1880–1964* (Boston, Mass.: Little, Brown & Company, 1978).
23. See John E. Endicott and William R. Heaton, *The Politics of East Asia* (Westview Press, 1978) p. 145.
24. Conversation with Amaya Naohiro, the then Deputy Vice Minister of International Trade and Industry, currently serving as the Executive Director of Dentsu Institute for Human Studies.
25. My first encounter with it was May 1986.
26. *The Atlanta Journal-Constitution*, 29 March 1992.
27. Ibid.
28. Lecture, Atlanta, Georgia, from Dentsu Institute of Human Studies, October 1991.
29. *The Daily Japan Digest*, 6 April 1992.
30. *The Japan Times*, 8 February 1992. Some 9982 males and 1279 females took the November 1991 exams for entrance into the Defense Academy. A total of 1192 males passed and 59 females. An additional 13 females had already passed an exam offered in September 1991, so that the grand total of females to enter the college will be 72 according to this newspaper account.

31. *Nihon Keizai*, 4 February 1992.
32. *The Daily Japan Digest*, 22 March 1992.
33. *Yomiuri Shimbun*, 19 December 1991.
34. Ibid, 30 January 1992.

Index